高等学校专业英语教材

# 电子技术专业英语教程
## （第 3 版）

冯新宇　寇晓静　陈晓洁　主编

电子工业出版社
Publishing House of Electronics Industry
北京·BEIJING

## 内 容 简 介

本书从电子类相关专业入手,把电子、信息、电气、自动化、测控等相关专业的重点内容进行了整合,涵盖了电子及相关专业领域的主要分支。全书由 16 个单元组成,分别是英语翻译技巧、电子基础、半导体元器件、运算放大器、电源、电子仪器、线性电路分析、数字逻辑电路、集成电路、微型计算机、程序设计语言、通信网络技术、数字多媒体系统、电子系统、EDA 工具、IC 手册。从第 2 单元开始,每个单元都有几个主题,每个主题都包含课文、词汇、注释和练习。其中,课文着力体现了该单元的核心关键技术、国外优秀科技成果、前沿领域及未来前景等;练习包括 Keywords 和 Summary 两部分,培养学生归纳总结的能力。本书网上教学资源包括电子教案、授课建议、参考译文、练习参考答案及科技英语写作要求,读者可登录华信教育资源网 www.hxedu.com.cn 免费注册下载。

本书可作为高等院校电子技术等相关专业的专业英语教材,也可供从事相关专业的工程技术人员学习参考。

未经许可,不得以任何方式复制或抄袭本书之部分或全部内容。
版权所有,侵权必究。

图书在版编目(CIP)数据

电子技术专业英语教程 / 冯新宇,寇晓静,陈晓洁主编. —3 版. —北京:电子工业出版社,2018.6
高等学校专业英语教材
ISBN 978-7-121-34235-6

I. ①电… II. ①冯… ②寇… ③陈… III. ①电子技术-英语-高等学校-教材 IV. ①TN

中国版本图书馆 CIP 数据核字(2018)第 106130 号

策划编辑:秦淑灵
责任编辑:苏颖杰
印　　刷:北京虎彩文化传播有限公司
装　　订:北京虎彩文化传播有限公司
出版发行:电子工业出版社
　　　　　北京市海淀区万寿路 173 信箱　邮编　100036
开　　本:787×1092　1/16　印张:15.25　字数:508 千字
版　　次:2009 年 8 月第 1 版
　　　　　2018 年 6 月第 3 版
印　　次:2024 年 7 月第 10 次印刷
定　　价:45.00 元

凡所购买电子工业出版社图书有缺损问题,请向购买书店调换。若书店售缺,请与本社发行部联系,联系及邮购电话:(010)88254888,88258888。
质量投诉请发邮件至 zlts@phei.com.cn,盗版侵权举报请发邮件至 dbqq@phei.com.cn。
本书咨询联系方式:qinshl@phei.com.cn。

# 前　言

专业英语教学的主要目的是指导学生阅读本专业及相关领域的英语书籍、文献，提高科技英语写作能力，使学生能以英语为工具获取相关的专业信息。本书根据第2版出版以来近4年的教学反馈再次进行修订而成，对过时的内容进行了删减，并增加了新的内容。新增加的内容除考虑了内容的时效性外，还考虑了专业英语授课的特殊性，借鉴同行优秀的教学方法，每个单元最后都增加了专业英语知识，书中原有经典章节继续保留，调整后的内容更适合本科层次专业英语教学。主要修订内容如下：删除 Reading Materials 部分共28篇课文，因为这部分内容在教学过程中使用不多，现在将其作为电子资料发布在互联网上，感兴趣的师生可以自行下载查看；开篇加入4个学时科技英语的翻译技巧，这些技巧在专业英语的授课中使用频率较高，老师反响也较好；从第2单元开始，每个单元都增加了"科技英语知识"，给枯燥的专业英语授课提供了素材，也达到了专业英语学习的目的，主要内容有数学符号及数学表达式的读法、形容词的翻译、英文书信、语法成分的转换、长句的翻译、专业英语词汇特点、专业英语概述、数词的翻译、名词的翻译、论文的标题和摘要、科技英语翻译标准、科技论文的结构与写作、动词的翻译、被动句的翻译、说明书常用术语等。

本书仍由16个单元组成，分别是英语翻译技巧、电子基础、半导体元器件、运算放大器、电源、电子仪器、线性电路分析、数字逻辑电路、集成电路、微型计算机、程序设计语言、通信网络技术、数字多媒体系统、电子系统、EDA 工具、IC 手册。从第2单元开始，每个单元都有几个主题，每个主题都包含课文、课文词汇、注释和练习。其中，课文着力体现了该单元的核心关键技术、国外优秀科技成果、前沿领域及未来前景等；练习包括 Keyword 和 Summary 两部分，培养学生归纳总结的能力。

本书由冯新宇主持制订编写大纲，对全书进行统稿，冯新宇、寇晓静、陈晓洁担任主编。其中，Unit 1 由王蕴恒编写，Unit 2～4 由陈晓洁编写，Unit 5～8 由蒋洪波编写，Unit 9～10 及15个单元的科技英语知识点由寇晓静编写，Unit 11～13 由冯新宇编写，Unit 14～16 单元由王娟编写。

在本书编写过程中，哈尔滨工业大学的张鹏楠、孙宇和黑龙江科技大学夏洪洋等老师对书稿的翻译进行了校准工作，在此表示感谢！

由于水平所限，书中难免有纰漏和欠妥之处，望读者不吝赐教。如有反馈意见，请发电子邮件至 123244441@qq.com。

编　者

# Contents

Unit 1　English-Chinese Translation
　　　　Techniques …………………… (1)
　Lesson 1　Morphology in English-Chinese
　　　　　　Translation …………… (1)
　Lesson 2　Syntax in English-Chinese
　　　　　　Translation …………… (9)
Unit 2　Introduction ………………… (17)
　Lesson 3　Semiconductor Materials … (17)
　　New Words ……………………… (19)
　　Phrases & Expressions ………… (20)
　　Technical Terms ………………… (20)
　　Notes ……………………………… (20)
　Lesson 4　Moore's Law …………… (20)
　　New Words ……………………… (23)
　　Phrases & Expressions ………… (23)
　　Technical Terms ………………… (23)
　　Notes ……………………………… (24)
　Exercises …………………………… (24)
　科技英语知识1：数学符号及数学
　　表达式的读法 …………………… (25)
Unit 3　Semiconductor Device ……… (26)
　Lesson 5　Resistors, Capacitors and
　　　　　　Inductors ……………… (26)
　　New Words ……………………… (29)
　　Phrases & Expressions ………… (29)
　　Technical Terms ………………… (29)
　　Notes ……………………………… (29)
　Lesson 6　Diode …………………… (30)
　　New Words ……………………… (33)
　　Phrases & Expressions ………… (33)
　　Technical Terms ………………… (33)
　　Notes ……………………………… (34)
　Lesson 7　Transistor ……………… (35)
　　New Words ……………………… (39)
　　Phrases & Expressions ………… (40)

　　Technical Terms ………………… (40)
　　Notes ……………………………… (40)
　Exercises …………………………… (41)
　科技英语知识2：形容词的翻译 …… (42)
Unit 4　Operational Amplifier ……… (44)
　Lesson 8　Performance Parameters … (44)
　　New Words ……………………… (46)
　　Phrases & Expressions ………… (46)
　　Technical Terms ………………… (46)
　　Notes ……………………………… (46)
　Lesson 9　Operational Amplifiers … (47)
　　New Words ……………………… (51)
　　Phrases & Expressions ………… (52)
　　Technical Terms ………………… (52)
　　Notes ……………………………… (52)
　Lesson 10　Op-amp Applications … (52)
　　New Words ……………………… (54)
　　Phrases & Expressions ………… (55)
　　Technical Terms ………………… (55)
　　Notes ……………………………… (55)
　Exercises …………………………… (55)
　科技英语知识3：英文书信 ………… (56)
Unit 5　Power ………………………… (57)
　Lesson 11　Buck-Boost Power Stage
　　　　　　　and Steady-State
　　　　　　　Analysis ……………… (57)
　　New Words ……………………… (62)
　　Phrases & Expressions ………… (62)
　　Technical Terms ………………… (62)
　　Notes ……………………………… (63)
　Lesson 12　Buck-Boost Steady-State
　　　　　　　Discontinuous Conduction
　　　　　　　Mode Analysis ……… (64)
　　New Words ……………………… (67)
　　Phrases & Expressions ………… (67)

  Technical Terms ............ (67)
  Notes ........................... (67)
 Lesson 13 Flyback Power Stage ... (68)
  New Words ..................... (71)
  Phrases & Expressions ......... (71)
  Technical Terms ............... (71)
  Notes .......................... (71)
 Exercises ......................... (72)
 科技英语知识 4：语法成分的转换 ... (73)
Unit 6 Electronic Instruments ........ (74)
 Lesson 14 Oscilloscope ......... (74)
  New Words ..................... (76)
  Phrases & Expressions ......... (76)
  Technical Terms ............... (76)
  Notes .......................... (77)
 Lesson 15 Function Generators .... (77)
  New Words ..................... (81)
  Phrases & Expressions ......... (81)
  Technical Terms ............... (82)
  Notes .......................... (82)
 Lesson 16 Computer-Based Test
     Instruments ......... (82)
  New Words ..................... (87)
  Phrases & Expressions ......... (87)
  Technical Terms ............... (87)
  Notes .......................... (87)
 Exercises ......................... (88)
 科技英语知识 5：长句的翻译 ......... (89)
Unit 7 Linear Circuit Analysis ........ (91)
 Lesson 17 Ohm's Law ............ (91)
  New Words ..................... (94)
  Phrases & Expressions ......... (95)
  Technical Terms ............... (95)
  Note ........................... (95)
 Lesson 18 Kirchhoff's Laws ...... (95)
  New Words ..................... (100)
  Phrases & Expressions ......... (100)
  Technical Terms ............... (100)
  Notes .......................... (101)
 Lesson 19 Circuit Analysis
     Methods ............ (101)

  New Words ..................... (105)
  Phrases & Expressions ......... (106)
  Technical Terms ............... (106)
 Exercises ......................... (106)
 科技英语知识 6：专业英语词汇
  特点 ............................ (107)
Unit 8 Digital Logic Circuit ........... (108)
 Lesson 20 Digital Systems ....... (108)
  New Words ..................... (109)
  Phrases & Expressions ......... (109)
  Technical Terms ............... (109)
  Note ........................... (110)
 Lesson 21 Logic Gates .......... (110)
  New Words ..................... (114)
  Phrases & Expressions ......... (114)
  Technical Terms ............... (114)
 Lesson 22 Flip-flop ............. (114)
  New Words ..................... (118)
  Phrases & Expressions ......... (118)
  Technical Terms ............... (118)
  Notes .......................... (118)
 Exercises ......................... (119)
 科技英语知识 7：专业英语概述 ... (120)
Unit 9 Integrated Circuits ............ (122)
 Lesson 23 Wafers ............... (122)
  New Words ..................... (124)
  Phrases & Expressions ......... (124)
  Technical Terms ............... (124)
  Notes .......................... (125)
 Lesson 24 IC Processing
     Technology ......... (125)
  New Words ..................... (128)
  Phrases & Expressions ......... (128)
  Technical Terms ............... (128)
  Notes .......................... (129)
 Lesson 25 IC Design Flow ....... (129)
  New Words ..................... (132)
  Phrases & Expressions ......... (132)
  Technical Terms ............... (132)
  Note ........................... (133)
 Exercises ......................... (133)

科技英语知识 8：数词的翻译 …… (134)

## Unit 10　Microcomputer …………… (136)
Lesson 26　Universal Serial Bus … (136)
　　New Words ………………………… (137)
　　Phrases & Expressions ………… (138)
　　Notes ……………………………… (138)
Lesson 27　MCS-51 ………………… (138)
　　New Words ………………………… (142)
　　Phrases & Expressions ………… (142)
　　Technical Terms ………………… (142)
Lesson 28　ARM …………………… (142)
　　New Words ………………………… (146)
　　Phrases & Expressions ………… (146)
　　Technical Terms ………………… (146)
　　Notes ……………………………… (146)
Exercises ………………………………… (147)
科技英语知识 9：名词的翻译 …… (148)

## Unit 11　Computer Program Design … (149)
Lesson 29　Introduction to C …… (149)
　　New Words ………………………… (151)
　　Technical Terms ………………… (151)
　　Notes ……………………………… (152)
Lesson 30　HDL Language ……… (152)
　　New Words ………………………… (155)
　　Phrases & Expressions ………… (155)
　　Technical Terms ………………… (155)
　　Notes ……………………………… (156)
Lesson 31　Perl …………………… (156)
　　New Words ………………………… (158)
　　Phrases & Expressions ………… (159)
　　Technical Terms ………………… (159)
　　Note ……………………………… (160)
Exercises ………………………………… (160)
科技英语知识 10：论文的标题
　　和摘要 ……………………………… (161)

## Unit 12　Communication Network
　　Technology ………………………… (162)
Lesson 32　Basic Telecommunications
　　Network ……………………… (162)
　　New Words ………………………… (164)
　　Phrases & Expressions ………… (164)
　　Technical Terms ………………… (164)
　　Note ……………………………… (164)
Lesson 33　Third Generation Cellular
　　Systems ……………………… (164)
　　New Words ………………………… (166)
　　Technical Terms ………………… (166)
　　Notes ……………………………… (166)
Lesson 34　Multimedia Network … (167)
　　New Words ………………………… (169)
　　Phrases & Expressions ………… (169)
　　Technical Terms ………………… (169)
　　Notes ……………………………… (169)
Exercises ………………………………… (170)
科技英语知识 11：科技英语翻译
　　标准 ……………………………… (171)

## Unit 13　Digital Multimedia Systems … (172)
Lesson 35　Basic Concepts in Digital
　　Multimedia Systems … (172)
　　New Words ………………………… (173)
　　Phrases & Expressions ………… (174)
　　Technical Terms ………………… (174)
　　Note ……………………………… (174)
Lesson 36　DVD …………………… (174)
　　New Words ………………………… (176)
　　Phrases & Expressions ………… (176)
　　Technical Terms ………………… (176)
　　Note ……………………………… (176)
Lesson 37　Compression
　　Methods ……………………… (176)
　　New Words ………………………… (180)
　　Phrases & Expressions ………… (180)
　　Technical Terms ………………… (180)
　　Notes ……………………………… (181)
Exercises ………………………………… (182)
科技英语知识 12：科技论文的
　　结构与写作 ……………………… (183)

## Unit 14　Electric Systems …………… (185)
Lesson 38　Personal Computer … (185)
　　New Words ………………………… (188)
　　Phrases & Expressions ………… (188)

  Technical Terms ·············· (189)
  Notes ························· (189)
 Lesson 39 Advanced Automated
      Fingerprint Identification
      System ··············· (190)
  New Words ··················· (193)
  Phrases & Expressions ········ (193)
  Technical Terms ·············· (194)
  Notes ························· (194)
 Lesson 40 System on a Programmable
      Chip ·················· (194)
  New Words ··················· (196)
  Technical Terms ·············· (196)
  Notes ························· (197)
 Exercises ······················· (197)
 科技英语知识 13：动词的翻译 ······ (198)
Unit 15 EDA Tools ··············· (199)
 Lesson 41 MATLAB ············· (199)
  New Words ··················· (202)
  Phrases & Expressions ········ (202)
  Technical Terms ·············· (202)
  Notes ························· (202)
 Lesson 42 PSpice ··············· (203)
  New Words ··················· (206)
  Phrases & Expressions ········ (206)

  Technical Terms ·············· (206)
  Notes ························· (207)
 Lesson 43 Cadence ············· (207)
  New Words ··················· (209)
  Phrases & Expressions ········ (209)
  Technical Terms ·············· (209)
  Notes ························· (209)
 Exercises ······················· (210)
 科技英语知识 14：被动句的翻译 ····· (211)
Unit 16 IC Manuals ··············· (212)
 Lesson 44 IC Datasheet ·········· (212)
  New Words ··················· (215)
  Phrases & Expressions ········ (215)
  Technical Terms ·············· (215)
  Notes ························· (215)
 Lesson 45 D Flip-flop ············ (216)
 Exercises ······················· (218)
 科技英语知识 15：说明书常用
   术语 ······················· (219)
**附录** ····························· (221)
 附录 A Technical Terms ········ (221)
 附录 B Scientific English
    Writing ················· (226)
**参考文献** ·························· (232)

# Unit 1　English-Chinese Translation Techniques

## Lesson 1　Morphology in English-Chinese Translation

### 1. Differences between English and Chinese

English is a tree-like sentence structure language. An English sentence has a trunk bringing forth all the branches. The complexity of a sentence does not affect its trunk.

Example 1. **The use of heat pumps has been held back** largely by skepticism about advertiser's claims that heat pumps can provide as many as two units of thermal energy for each unit of electrical energy used, thus apparently contradicting the principle of energy conservation.

译文：**热泵的使用受到阻碍**主要是由于消费者对广告商宣称热泵消耗一个单位的电能可以提供两个单位的热能这种说法持怀疑态度而引起的，因为这显然同能量守恒原理相矛盾。

Example 2. **The computer performs a supervisory function** in the liquid-level control system by analyzing the process conditions against desired performance criteria and determining the changes in process variables to achieve optimum operation.

译文：在液位控制系统中，**计算机执行着一种监控功能**，它根据给定的特性指标来分析各种过程条件，并决定各过程变量的变化以获得最佳操纵。

Example 3. **There is no more difference**, but there is just the same kind of difference, between the mental operations of a man of science and those of an ordinary person, as there is between the operations and methods of a baker or of a butcher weighing out his goods in common scales, and the operations of a chemist in performing a difficult and complex analysis by means of his balance and finely graded weights.

译文：(没有什么更大的差别)科研人员的思考过程与普通人的思考过程的差别就等同于面包师或屠夫用普通的秤称量商品与化学家用天平和精密砝码进行困难复杂的分析之间的差别。

A Chinese sentence has a bamboo-like structure with different sections linking one another, each one expressing a relatively complete semantic meaning independently.

Example 4. "碧云天，黄叶地，西风紧，北雁南飞。"

English grammar is expressed by conjunctions, relative pronouns and morphological changes. In Chinese, a variety of grammatical relations are implied in the context such as active and passive, tense.

Example 5. 我以为你回家了。I thought you had gone home.

Example 6. 工作做完了。The work has been finished.

### 2. Methods in English-Chinese translation

As a general criterion of translation, the translators should not add or reduce text content at will. However, due to the big difference between English and Chinese, it is difficult to find corresponding word in the new text during the actual translation process. Therefore, in order to accurately express the original information, the translators often need to do some addition or subtraction of translation. So some people think, a good translation is generally slightly longer than the original — because the original author and the SL readers rarely have the language and cultural barriers, but the TL reader does not have such convenience, so translators often need to add the content implied in the original for the TL reader to understand easily.

1) Amplification in English-Chinese translation

Amplification in English-Chinese Translation is mainly for the consideration of Chinese expression, which includes amplification of words omitted from the original and adding the necessary connectives, quantifiers, or plural concept, or for the considered of rhetorical coherence, so that translation can conform to Chinese idiomatic expression.

(1) Amplification by supplying words omitted in the original

Example 7. Matter can be changed into energy, and energy into matter.

译文：物质可以转化为能，能也可以**转化**为物质。

Example 8. The best conductor has the least resistance and the poorest the greatest.

译文：最好的导体电阻最小，最差的**导体**电阻最大。

Example 9. A proton has a positive charge and an electron a negative charge, but a neutron has neither.

译文：质子带正电，电子带负电，而中子**既不带正电也不带负电**。

(2) Amplification by supplying necessary connectives

Example 10. Heated, water will change into vapor.

译文：水如受热，**就会**汽化。

Example 11. However carefully boiler casings and steam pipes are sealed, some heat escapes and is lost.

译文：不管锅炉壳与蒸汽管封闭得多么严密，**还是**有一部分热量散失并损耗掉。

Example 12. Since air has weight, it exerts force on any object immersed in it.

译文：因为空气具有重量，**所以**处在空气中的任何物体都会受到空气的作用力。

Example 13. The pointer of ampere meter moves from zero to five and goes back to two.

译文：安培表的指针**先**从 0 转到 5，**然后**又回到 2。

(3) Amplification by supplying words to convey the concept of plurality

Example 14. Note that the **words** "velocity" and "speed" require explanation.

译文："速度"和"速率"这**两个词**需要解释。

Example 15. He stretched his **legs** which were scattered with **scars**.

译文：他伸出**双腿**，露出腿上的道道**伤痕**。

Example 16. The mountains began to throw their long blue **shadows** over the valley.

译文：群山开始向山谷投下一道道蔚蓝色**长影**。

(4) Amplification by supplying words to make an abstract concept clear

Example 17. **Oxidation** will make iron and steel rusty.

译文：**氧化作用**会使钢铁生锈。

Example 18. **This lack of resistance** in very cold metals may become useful in electronic computers.

译文：这种在甚低温金属中**没有电阻的现象**可能对电子计算机很有用处。

Example 19. After all **preparations** were made, the plane took off.

译文：一切**准备工作**就绪以后，飞机就起飞了。

(5) Amplification by logical thinking

Example 20. Air pressure **decreases** with altitude.

译文：气压随海拔的**增加**而下降。

Example 21. I was taught that **two sides of a triangle** were greater than the third.

译文：我学过，三角形的**两边之和**大于第三边。

Example 22. This shows that the resistance of an electric conductor is inversely proportional to its cross-section area.

译文：这表明，导体电阻的**大小**与导体横截面积的**大小**成反比。

(6) Amplification by supplying words of generalization

Example 23. According to scientists, it takes nature 500 years to create an inch of topsoil.

译文：根据科学家们的**看法**，自然界要用500年的时间才能形成一英寸厚的表层土壤。

Example 24. The thesis summed up the new achievements made in electronic computers, artificial satellites and rockets.

译文：该论文总结了电子计算机、人造卫星和火箭**三方面**的新成就。

Example 25. The principal functions that may be performed by vacuum tubes are rectification, amplification, oscillation, modulation, and detection.

译文：真空管的**五大**主要功能是整流、放大、振荡、调制和检波。

(7) Amplification by necessary repetition

Example 26. Avoid using this computer in **extreme** cold, heat, dust or humidity.

译文：**过冷、过热、灰尘过重、湿度过大**的情况下，不要使用此计算机。

Example 27. I had experienced oxygen and/or engine **trouble**.

译文：我曾经遇到的情况，不是氧气设备**出故障**，就是引擎**出故障**，或者两者皆有。

(8) Amplification by necessary modification

Example 28. They could sense **her (Titanic) mass**, her eerie quiet and ruined splendor of a lost age.

译文：他们体会到它（泰坦尼克号）**船体的巨大**，感受到它周围的可怕沉寂，领略到因岁月的流逝而造成的凄凉景观。

2) Omission in English-Chinese translation

In general, Chinese is more concise than English. In English-Chinese translation, if many indispensable words in the original are translated into Chinese word for word, they will become unnecessary verbiage. Translation would seem rather cumbersome. Therefore omission is widely used in English-Chinese translation. Its main purpose is to delete some of the dispensable, unsuitable words for idiomatic expressions, such as omission of the substantives as pronouns, verbs and functional words as articles, prepositions, conjunctions, etc.

(1) Omission of the pronoun

Pronouns are more frequently used in English than in Chinese. Therefore, when translated into Chinese, many English pronouns may be omitted so as to conform the rendering to the accustomed usage of Chinese expression.

Example 29. Different metals differ in **their** conductivity.

译文：不同的金属导体的导电性不同。

Example 30. The current will blow the fuses when **it** reaches certain limit.

译文：当电流达到一定界限时会使熔丝熔断。

(2) Omission of the article

The article is the hallmark of English nouns. When translated into Chinese, it is usually omitted except when the indefinite article is intended to indicate the numeral "one", or "a certain".

Example 31. **The** controlled output is the process quantity being controlled.

译文：被控输出量是指被控的过程变量。

Example 32. Any substance is made up of atoms whether it is **a** solid, **a** liquid, or **a** gas.

译文：任何物质，不论是固体、液体还是气体，都由原子组成。

Example 33. The direction of **a** force can be represented by **an** arrow.

译文：力的方向可以用箭头表示。

(3) Omission of the preposition

Chinese is characterized by its succinctness and the preposition appears less frequently in Chinese than in English, and therefore omission of prepositions is a common practice in English-Chinese translation.

Example 34. The difference **between** the two machines consists in power.

译文：这两台机器的差别在于功率不同。

Example 35. Hydrogen is the lightest element **with** an atomic weight of 1.0008.

译文：氢是最轻的元素，原子量为1.0008。

(4) Omission of the conjunction

Chinese is considered an analytic language and many conjunctions that are indispensable in English may seem redundant in Chinese. Therefore, omission of the conjunction is a common practice in English-Chinese translation.

Example 36. **If** I had known it, I would not have joined in it.

译文：早知如此，我就不参加了。

Example 37. Like charges repel each other **while** opposite charges attract.

译文：同性电荷相斥，异性电荷相吸。

(5) Omission of the verb

As Chinese is a language of parataxis, its grammar is not so strict as that of Englsh, and predicative verbs in Chinese sometimes may also be omitted.

Example 38. When the pressure **gets** low, the boiling point **becomes** low.

译文:气压低,沸点就低。

Example 39. Solids expand and contract as liquids and gases **do**.

译文:如同液体和气体一样,固体也能膨胀和收缩。

Example 40. For this reason television signals **have** a short range.

译文:因此,电视信号的传播距离很短。

(6) Omission of the impersonal pronoun "it"

Example 41. This formula makes **it** easy to determine the wavelength of sounds.

译文:这一公式使得测定声音的波长十分简单。

Example 42. **It** was not until the middle of the 19th century that the blast furnace came into use.

译文:直到 19 世纪中叶,高炉才开始使用。

3) Conversion in English-Chinese translation

(1) Conversion into verb

One of the most remarkable differences between English and Chinese syntax is the use of the verb. It is taken for granted that an English sentence contains no more than one predicate verb, while in Chinese it is not unusual to have clusters of verbs in a simple sentence. Take the following sentence for example.

Example 43. Families upstairs have to carry pails to the hydrant downstairs for water.

译文:**住**在楼上的人家得**提**着水桶**去**楼下的水龙头处**打**水。

Four verbs are clustered in the Chinese version for an English simple sentence. This indicates obviously that in English-Chinese translation the conversion of English words of various parts of speech into Chinese verbs is a matter of common occurrence.

① Nouns converted into verbs

Example 44. Rockets have found **application** for the **exploration** of the universe.

译文:火箭**用**于**探索**宇宙。

Example 45. The **sight** and **sound** of our jet planes filled me with special longing.

译文:**看到**我们的喷气式飞机,**听见**隆隆的声音,我心驰神往。

② Prepositions converted into verbs

Example 46. A force is needed to move an object **against** inertia.

译文:为使物体**克服**惯性而运动,就需要一个力。

Example 47. Noise figure is minimized **by** a parameter amplifier.

译文:**采用**参数放大器可将噪声指数降到最低。

③ Adjectives converted into verbs

Example 48. Both of the substances are **soluble** in water.

译文:这两种物质都能**溶**于水。

Example 49. If low-cost power becomes **available** from nuclear power plants, the electricity crisis would be solved.

5

译文:如果能从核电站**获得**低成本电力,就会解决电力紧张问题。

④ Adverbs converted into verbs

Example 50. When the switch is **off**, the circuit is open and electricity doesn't go through.

译文:当开关**断开**时,电路就形成开路,电流不能通过。

Example 51. In this case the temperature in the furnace is **up**.

译文:这种情况下,炉温就**升高**。

(2) Conversion into nouns

Nouns account for an overwhelming part of the vocabulary both in Chinese and in English. Conversion between nouns and other parts of speech is also frequently adopted in English-Chinese translation.

① Verbs converted into nouns

Some verbs that are derived from nouns can hardly be translated literally so they are usually converted into nouns.

Example 52. Such materials are **characterized** by good insulation and high resistance to wear.

译文:这些材料的**特点**是:绝缘性好,耐磨性强。

Example 53. The design **aims** at automatic operation, easy regulation, simple maintenance and high productivity.

译文:设计的**目的**在于自动工作、调节方便、维护简易、生产率高。

② Adjectives converted into nouns

Adjectives with the definite articles to indicate categories of people, things, or adjectives used as predicative to indicate the nature of things may also be converted into nouns.

Example 54. Both the compounds are acids, **the former** is strong, **the latter** weak.

译文:这两种化合物都是酸,**前者**是强酸,**后者**是弱酸。

Example 55. In the fission process the fission fragments are very **radioactive**.

译文:在裂变过程中,裂变碎片具有强烈的**放射性**。

Example 56. IPC is more **reliable** than common computer.

译文:工控机的**可靠性**比普通计算机高。

③ Pronouns converted into nouns

English pronouns are more frequently used than Chinese pronouns. In order to make clear what they really refer to, we sometimes have to convert them into nouns, i. e., to repeat the nouns that they stand for.

Example 57. Radio waves are similar to light waves except that **their** wavelength is much greater.

译文:无线电波与光波相似,但**无线电波**的波长要长得多。

Example 58. The specific resistance of iron is not so small as **that** of copper.

译文:铁的电阻系数不如铜的**电阻系数**那样小。

Example 59. The result of this experiment is much better than **those** of previous ones.

译文:这次实验的结果比前几次的**实验结果**好得多。

Example 60. Experiment indicates that the new chip is about 2 times as integrative as **that** of the old ones.

译文:实验表明新型芯片的集成度是旧型芯片**集成度**的 2 倍。

(3) Conversion into adjectives

In English-Chinese translation there are some circumstances in which various other parts of speech in English can be converted into Chinese adjectives.

① Nouns converted into adjectives

Example 61. This experiment was a **success**.

译文:这个实验很**成功**。

② Adverbs converted into adjectives

Example 62. Earthquakes are **closely** related to faulting.

译文:地震与地层断裂有**密切的**关系。

Example 63. It is demonstrated that gases are **perfectly** elastic.

译文:人们已经证实,气体具有**理想的**弹性。

Example 64. The pressure **inside** equals the pressure **outside**.

译文:**内部的**压力和**外部的**压力相等。

(4) Conversion into adverbs

Sometimes, for the sake of convenience, some parts of speech in English may be converted into Chinese adverbs in English-Chinese translation.

Example 65. Below 4℃, water is in **continuous** expansion instead of **continuous** contraction.

译文:水在 4℃以下就会**不断地**膨胀,而不是**不断地**收缩。

Example 66. Rapid evaporation at the heating-surface **tends** to make the steam wet.

译文:加热面上的迅速蒸发,**往往**使水蒸气的湿度变大。

Example 67. The **same** principles of low internal resistance also apply to milliammeters.

译文:低内阻原理也**同样**适用于毫安表。

(5) Conversion of sentence elements

Conversion in a broader sense includes the conversion of the "active voice" into the "passive". Sometimes, it may involve the change of various elements of a sentence, such as from the subject to the object, and vice versa.

Example 68. As the match burns, **heat and light** are given off. (from the subject to the object)

译文:火柴燃烧时发出**光和热**。(从主语到宾语)

Example 69. This sort of stone has a **relative** density of 2.7. (from the object to the subject)

译文:这种石头的**相对密度**是 2.7。(从宾语到主语)

Example 70. **Care** must be taken at all times to protect the instrument from dust and damp. (from the subject to the predicate)

译文：应当始终**注意**保护仪器，不使其沾染灰尘和受潮。（从主语到谓语）

Example 71. Fuzzy control **is found** a effective way to control the systems without precise mathematic models.

译文：**人们发现**，模糊控制是一种控制不具备精确数学模型系统的有效方法。（从被动到主动）

4) Restructuring in English-Chinese translation

(1) Different sequences in customary word combinations

Every nation has its own unique ways of combining words, and so are the English and Chinese.

Example 72. Phase: heart-warming, tough-minded, East China, north and south, east and west, share the weal and woe, rain or shine, inconsistency of deeds with words, the iron and steel industry, quick of eye and deft of hand, food, clothing, shelter and transportation, you, he and I, back and forth, to and fro, every means possible.

译文：短语：暖人心的，意志坚定的，华东，东南西北，祸福与共，无论晴雨，言行不一，钢铁工业，手疾眼快，衣食住行，你我他，前前后后，来来去去，一切可能的手段。

Example 73. Connect the black pigtail with the dog-house.

译文：把黑色的猪尾巴系在狗窝上。（错误）

译文：将黑色的引出线接在高频高压电源屏蔽罩上。（正确）

Example 74. Outside it was pitch-dark and it was raining cats and dogs.

译文：外面一片漆黑，大雨倾盆。

(2) Different sequences in customary sentence arrangement

English sentences are generally bound by syntactical rules, and they are often arranged in order of importance, while Chinese sentences are more flexible, and usually arranged according to the sequence of time.

Example 75. Rocket research **has confirmed a strange fact** which had already been suspected: there is a "high-temperature belt" in the atmosphere, with its center roughly thirty miles above the ground.

译文：人们早就怀疑大气层中有一个"高温带"，其中心在距地面约30英里高的地方。利用火箭加以研究后，**这一奇异的事实已得到了证实**。

Example 76. **We had been dismayed** at home while reading of the natural calamities that followed one another for three years after we left China in 1959.

译文：我们于1959年离开了中国。此后，中国连续三年遭到自然灾害。当我们在国外读到这方面的消息时，**心情颇为沉重**。

(3) Adverbial clauses in a complex sentence

Adverbial clauses of condition, purpose, concession or cause, and so on in English may stand either before or after a principal clause, however, when translated into Chinese, they are normally placed before the principal clause.

Example 77. The government is determined to keep up the pressure **whatever the cost it will pay in the end.**

译文：**不论最终将付出什么代价**，政府都决心继续施加压力。

Example 78. **No matter how hard she tried**, she failed at last.

译文：**不管她怎样努力**，最后还是没有成功。

(4) "That-" clause used as a judgment or conclusion

"That" clause is usually adopted in expressing a judgment or conclusion. When a "that" clause which serves as the subject (of a sentence or a clause) is too long and consequently "it" is simultaneously used as the "formal subject", restructuring is often called for, so as to conform to the Chinese usage.

Example 79. **It is not surprising that**, when humidity is low, the water evaporates rapidly from the fruit.

译文：在大气湿度低的情况下，水果里的水分蒸发就快，**这是不足为奇的**。

Example 80. **It is common practice that** electric wires are made from copper.

译文：电线是铜制的，**这是常见的做法**。

Example 81. A few years ago **it was thought unbelievable that** the computer could have so high speed as well so small volume.

译文：几年前人们还认为计算机能具有如此高的运行速度和如此小的体积是**一件难以置信的事**。

# Lesson 2  Syntax in English-Chinese Translation

Due to extensive use of attributive clause, adverbial clause and a variety of phrases, English sentences often have longer complex structures. In addition, because of the differences between English and Chinese in the morphology, syntax, logical thinking and so on, it is difficult to understand and translate long sentences.

1. Translation techniques

The tree-like sentence structure of English is converted into the Chinese bamboo-like sentence structure: a long sentence is divided into pieces, so some necessary changes are needed to make them relatively independent and complete. A few relatively independent sentences are rearranged in accordance with Chinese custom, that is in the chronological order, then in the logical order from the conditions to the results.

Example 1. The newly described languages were often so strikingly different from the well studied languages of Europe and Southeast Asia that some scholars even accused Boas and Sapir of fabricating their data.

译文：这些新描述的语言与人们充分研究过的欧洲与东南亚的语言截然不同，以至于有些学者指责 Boas 与 Sapir 伪造资料。

Example 2. No such limitation is placed on an AC motor; here the only requirement is relative motion, and since a stationary armature and a rotating field system have numerous advantages, this arrangement is standard practice for all synchronous motor rated above a few kilovolt-amperes.

译文：交流电机不受这种限制，唯一的要求是相对运动，而且由于固定电枢及旋转磁场系统具有很多优点，所以这种安排是所有容量在几千伏安以上的同步电机的标准做法。

Example 3. The resistance of any length of a conducting wire is easily measured by finding the potential difference in volts between its ends when a known current is following.

译文：已知导线中流过的电流，只要测出导线两端电位差的伏值，就能很容易地得出任何长度导线的电阻值。

(1) Attributive clause

When attributive clause is a short or restrictive attributive clause, in general, it will be removed to the front and become a pre-modifier. When the attributive clause is longer, you can restore the antecedent, so that the attributive clause will become an independent sentence. In many cases, in accordance with the logic, in fact, the attributive clause is acted as adverbial, which indicates condition or cause, then, it should be translated into an adverbial clause.

Example 4. Each of its collectors placed on the roof consists of a shallow fibreglass tray **that holds a number of copper tubes under a special glass cover.**

译文：每个安放在屋顶的收集器都由一个浅浅的玻璃纤维盘组成，**上面是玻璃罩，罩下面有许多铜管。**

Example 5. The failure to act could mean untold misery for future generations and perhaps environmental disaster **which threaten our very existence.**

译文：如果不采取行动，就意味着对我们的子孙后代带来难以形容的苦难和**威胁我们生存**的环境灾难。

Example 6. No one can be a great thinker **who does not realize that as a thinker it is his first duty to follow his intellect to whatever conclusions it may lead.**

译文：如果你没有意识到作为思想家首要的职责是听从自己的理性而无论结果如何，那么你就不能成为伟大的思想家。

Example 7. We must combat the medical-psychiatric model of human behavior **that seeks a drug for every psychological discomfort** and **under which a person who is not continuously calm, anxiety-free, happy and content is define as a medical patient.**

译文：我们必须对抗这种人类行为的医学生理分析模式，它为每个有心理疾病的人寻求药物。这种模式下，任何不能持续平静、无忧无虑、幸福和心理满足的人都被定义为医学上的病人。（显然这种模式是不合理的，应注意理解句意的重要性。）

Example 8. We are not conscious of the extent to **which work provides the psychological satisfaction that can make the difference between a full and an empty life.**

译文：我们不清楚在多大程度上工作能够给我们带来满足感，从而使我们感到生活是充实的。（对工作能不能带来满足感持怀疑态度。）

Example 9. This kind of two-electrodes tube consists of a tungsten filament, **which gives off electrons when it is heated**, and a plate **toward which the electrons migrate when the field is in the right direction.**

译文:这种二极管由一根钨丝和一个极板组成。钨丝受热时放出电子,当电场方向为正时,电子就向极板移动。

(2) Appositive clause

There is no appositive clause in Chinese, and therefore it is difficult to translate English appositive clause into Chinese without any change. The general approach is to change words like "belief" into the verb "believe", and the appositive clause would become the object clause.

Example 10. This is due largely to the fact **that many writers think, not before, but as they write.**

译文:这主要是由于许多作者不是在写之前想,而是边写边想。

Example 11. Predictions **that the phenomenon of globalization will result in a lowering of human rights standards as the mobility of capital seeks out the markets least constrained by labor and human rights standards to maximize the highest returns** need not be the case.

译文:有人预测全球化现象将会导致人权标准降低,因为流动资本流向受劳工和人权标准束缚最小的市场以获取最大回报,但这种预测未必正确。

(3) Double negative

Typical translation: the affirmative structure

Example 12. **Until** humans use a spear to hunt game or a robot to produce machine parts, **neither** is much **more than** a solid mass of matter.

译文:如果人类不用矛去捕猎或者不用机器人去生产机器零件,那么它们只不过是一块废铁。

Example 13. **Few** technological developments have had a greater impact on our lives **than** the computer revolution.

译文:对我们生活影响最大的技术成果就是计算机革命。

Example 14. There are probably **no** questions we can think up that **can't** be answered, sooner or later, including even the matter of consciousness.

译文:凡是我们能够想到的问题迟早都会得到解决,甚至包括意识问题。

Example 15. There is **nothing** that can be said by mathematical symbols and relations which **canno**t also be said by words. The converse, however, is false. Much that can be and is said by words cannot successfully be put into equations, because it is nonsense.

译文:任何能用数学符号或数学关系式描述的都可以用语言描述,反过来是错的。很多能用语言表达的却不能转化为方程,因为这些语言是没有意义的。(讽刺语言中充满了废话。)

(4) Transferred negation

In English, "not" generally negates any part in its rear, that is termed negative transference.

Example 16. I did **not** have lunch **in the restaurant.**

译文:我在餐馆没吃午饭。(错误)

译文:我没在这家餐馆吃午饭。(正确,暗示我吃午饭了。)

Example 17. The computer is **not** valuable **because** it is expensive.

译文:计算机因为其价格贵而没有价值。(错误)

译文:计算机不是因为价格贵才有价值。(正确)

Example 18. The engine **didn't** stop **because** the fuel was finished.

译文:因为油用光了,所以引擎没停止工作。(错误)

译文:引擎停下了不是因为油用光了(正确,暗示可能是发动机坏了。)

Example 19. These cultural conventions are **no** less confining simply **because** they cannot be seen or touched.

译文:因为这些文化习俗看不见、摸不着,所以更多地束缚了人们。(错误)

译文:这些文化习俗不会仅仅因为看不见、摸不着而减小对人们的束缚。(正确)

Example 20. The importance of computer in the use of automatic control **can not** be **overestimated.**

译文:计算机在自动控制应用上的重要性不能被估计过高。(错误)

译文:对计算机在自动控制应用上的重要性怎么估计也不会过高。(正确)

(5) Inversion

Example 21. **Not until** these fundamental subjects were sufficiently advanced **was it** possible to solve the main problems of flight mechanics.

译文:只有当这些基础学科充分地发展完善之后,才能解决飞行力学的主要问题。

Example 22. **Into** this area of industry **came millions of Europeans** who made of it what become know as the "melting pot", the fusion of people from many nations into Americans.

译文:数百万欧洲人涌入这一工业地区,这些人使它成为"大熔炉",来自很多国家的人最后成为了美国人。

(6) Object postposition

In general, the object in English follows the verb closely. However, if the object is relatively long, the object complement or adverbial will be preposed and the object will be postposed, which will be difficult for us to understand. But the solution is pretty simple, just try to find out whether the verb is transitive. If it is a transitive verb, there must be an object. If the object doesn't follow the verb closely, try to locate it afterwards.

Example 23. Others do not try to widen their experience because they prefer the old and familiar, **seek** from their affairs **only further confirmation of the correctness of their own values.**

译文:其他人不想拓宽他们的经验,因为他们更喜欢旧的、熟悉的东西,以从中寻求进一步确认他们自己的价值观的正确性。

(7) Nominalization structure

This is a special kind of noun phrase. The important noun has the meaning of action, and therefore can be understood as a verb. Then the relationship between the following noun and the action is confirmed as either the recipient or sender of the action, and the other parts can be understood as adverbial.

Example 24. Coal represents **the chemical action of the sun** on the green plants thousands of years ago.

译文:煤代表的是几千年前**太阳**对绿色植物的**化学作用**。

Example 25. Coupled with the growing quantity of information is the development of technologies which enable **the storage and delivery of more information** with greater speed to more locations than has ever been possible before.

译文:伴随着信息量不断增加的是这些技术的发展,与过去相比,现在可以用更快的速度**储存和发送更多的信息**。

Example 26. The second aspect is the **application by** all members of society from the government official to the ordinary citizen, **of** the special methods of thought and action that scientists use in their work.

译文:问题的第二个方面是所有社会成员,从政府官员到普通百姓都去**运用**科学家们在工作中所用到的思想与行动方法。

Example 27. Continuous control readily lends itself to **an understanding of feedback control theory** using relatively uncomplicated mathematics.

译文:由于使用的数学原理相对简单,故连续控制有助于**理解反馈控制理论**。

(8) Partition structure

Two closely grammatically-related sentence elements are split by other sentence elements to form a partition structure.

Example 28. Ice particles **acquire a positive charge** bouncing off the falling hail.

译文:从冰雹弹开的冰粒获得了正电。

Example 29. The story has too many clues, **due to that it is written in stream of consciousness**, to be understood.

译文:这本小说是用意识流的手法写成的,因此线索太多,很难理解。

Example 30. In the early industrialized countries of Europe the process of industrialization **with all the far-reaching changes in social patterns that followed** was spread over nearly a century, whereas nowadays a developing nation may undergo the same process in a decade or so.

译文:在欧洲那些较早实现工业化的国家中,那种给后来的社会结构带来深远影响的工业化过程持续了100年,而如今一个发展中国家经历同样的过程很可能只用了10年左右的时间。

Example 31. Probably there is not one here who has not **in the course of the day** had occasion to set **in motion** a complex train of reasoning, of the very same kind, **though differing in degree**, as that which a scientific man goes through in tracing the causes of natural phenomena.

译文:很可能每个人每天都需要一系列的复杂思考,这种思考过程与科学家在探索自然现象的起因时所使用的思考过程相比,尽管程度不同,但性质是一样的。

(9) It clause

Example 32. **It doesn't** come as a surprise to you **to** realize that it makes no difference what you read or study if you can't remember it.

译文:如果你记不住的话,读什么和学什么就没有差别,这点对你来说毫不惊奇。

Example 33. Barnes suggested that **it was** as proper **to** term the plant a water structure as **to** call a house composed mainly of brick a brick building.

译文：Barnes 建议这种植物适合称为水结构，正如把主要由砖构成的房子称作砖混建筑一样。

Example 34. **It is** human, perhaps, **to** appreciate little that which we have and to long for that which we have not, but **it is** a great pity **that** in the world of light the gift of sight is used only as a mere convenience rather than as a means of adding fullness to life.

译文：不珍惜拥有的而渴望没有的东西，这是人之常情，但很遗憾的是，在光明的世界中，被赠予的光明仅仅被用作一种便利而非充实人生的方式。

Example 35. **It is** entirely reasonable for auditors **to** believe that scientists who know exactly where they are going and how they will get there should not be distracted by the necessity of keeping one eye on the cash register while the other eye is on the microscope.

译文：审计员认为既然科学家知道他们要达到什么目的，也知道如何达到这个目标，那么要求在搞好科研的同时要节省经费就不会使他们分心了，这种要求是合理的。

(10) More... than... structure

Notice: This sentence pattern is not comparative, but a contrast relationship is expressed. It can be translated into "not... but...". The emphasis is put on the content before "than".

Example 36. The book seems to be **more** a dictionary **than** a grammar.

译文：这本书**与其说**是语法书**不如说**是字典。

(11) No more... than... structure

Both of the words before and after "than" are negative, and it can be translated into neither... nor...

Example 37. I **no more** believe that ethics can be taught past the age of 16 **than** I believe in the teaching of so-called creative writing.

译文：**我不认为**人在16岁以后还能够学习伦理道德，**就像我不认为**人们能够学会创造性写作一样。

(12) not so much... as... structure

The sentence pattern denotes contrast that can be translated into "not... but...". The emphasis is put on the content after "as".

Example 38. The oceans do **not so much** divide the world **as** unite it.

译文：**与其说**海洋把世界分割开，**还不如说**海洋把世界连起来。

Example 39. Science moves forward, they say, **not so much** through the insights of great men of genius **as** because of more ordinary things like improved techniques and tools.

译文：他们说科学向前发展**不是**由于伟大天才的真知灼见，**而是**由于更普通的东西，比如不断改进的技术和工具。

Example 40. It is the admission of ignorance that leads to progress **not so much** because the solving of a particular puzzle leads directly to a new piece of understanding **as** because the puzzle if it interests enough scientists leads to work.

译文:承认无知才能走向进步,这么说**不是**因为解决某个特定的谜题导致一种新的理解,**而**在于这个谜题能够吸引科学家促使人们去研究。

(13) Elliptical sentence

Example 41. The breakthroughs would create as much danger as **they would hope.**

译文:这些突破与他们所希望的一样会带来危险。(错误)

译文:这些突破在**带来希望**的同时也带来危险。(正确)

Example 42. A whale is no more a fish than **a horse is.**

译文:鲸鱼不是鱼也不是马。(错误)

译文:鲸鱼不是鱼就像**马也不是鱼一样**明显。(正确)

Example 43. Away from their profession, scientists are inherently no more honest or ethical than **other people.**

译文:除他们的专业之外,从本质上说,科学家是不诚实、不道德的,就像其他人不诚实、不道德一样。(错误)

译文:在他们的专业之外,从本质上说,科学家并不比**普通人更诚实、更有道德**。(正确)

Example 44. Certainly the humanist thinkers of the eighteenth and nineteenth centuries, who are our ideological ancestors, thought that the goal of life was unfolding of a person's potentialities; what mattered to them was the person **who is much, not the one who has much or uses much.**

译文:18世纪和19世纪的人道主义思想家是我们意识形态上的先知,他们认为人生的目标是开发潜能。对他们来说,重要的是**他是什么样的人,而不是他拥有多少财富或使用了多少财富**。

2. Exercise

Quantum mechanics is a highly successful theory: it supplies methods for accurately calculating the results of diverse experiments, especially with minute particles. The predictions of quantum mechanics, however, give only the probability of an event, not a deterministic statement of whether or not the event will occur. Because of this probabilism, Einstein remained strongly dissatisfied with the theory throughout his life, though he did not maintain that quantum mechanics is wrong. Rather, he held that it is incomplete: in quantum mechanics the motion of a particle must be described in term of probabilities, he argued, only because some parameters that determine the motion have not been specified. If these hypothetical "hidden parameters" were known, a fully deterministic trajectory could be defined. Significantly, this hidden-parameter quantum theory leads to experimental predictions different from those of traditional quantum mechanics. Einstein's ideas have been tested by experiments performed since his death, and as most of these experiments support traditional quantum mechanics, Einstein's approach is almost certainly erroneous.

Echolocating bats emit sounds in patterns—characteristic of each species—that contain both frequency-modulated(FM) and constant-frequency(CF) signals. The broadband FM signals and the narrowband CF signals travel out to a target, reflect from it, and return to the hunting bat. In this process of transmission and reflection, the sounds are changed, and the changes in the echoes enable the bat to perceive features of the target. The FM signals report information about target characteristics that modify the timing and the fine frequency structure, or spectrum, of echoes — for example, the target's size, shape, texture, surface structure, and direction in space. Because of their narrow band-width, CF signals portray only the target's presence and, in the case of some bat species, its motion relative to the bat's. Responding to changes in the CF echo's frequency, bats of some species correct in flight for the direction and velocity of their moving prey.

# Unit 2 Introduction

## Lesson 3  Semiconductor Materials

In general, all materials may be classified in three major categories — conductors, semiconductors, and insulators — depending upon their ability to conduct an electric current. Semiconductor is a material that has a resistivity value between that of a conductor and an insulator. The conductivity of a semiconductor material can be varied under an external electrical field [1]. As the name indicates, a semiconductor material has poorer conductivity than a conductor, but better conductivity than an insulator.

The study of semiconductor materials began in the early 19th century. Over the years, many semiconductors have been investigated. The Table 3-1 shows a portion of the periodic table related to semiconductors. The elemental semiconductors are those composed of single species of atoms, such as silicon (Si), germanium (Ge), and gray tin (Sn) in column IV and selenium (Se) and tellurium (Te) in column VI. There are, however, numerous compound semiconductors that are composed of two or more elements. Gallium arsenide (GaAs), for example, is a binary III-V compound, which is a combination of gallium (Ga) from column III and arsenic (As) from column V.

Table 3-1  A portion of the periodic table showing elements used in semiconductor materials

| Period \ Group | II | III | IV | V | VI |
|---|---|---|---|---|---|
| 2 |  | B<br>Boron | C<br>Carbon | N<br>Nitrogen | O<br>Oxygen |
| 3 |  | Al<br>Aluminum | Si<br>Silicon | P<br>Phosphorus | S<br>Sulfur |
| 4 | Zn<br>Zinc | Ga<br>Gallium | Ge<br>Germanium | As<br>Arsenic | Se<br>Selenium |
| 5 | Cd<br>Cadmium | In<br>Indium | Sn<br>Tin | Sb<br>Antimony | Te<br>Tellurium |
| 6 | Hg<br>Mercury |  |  |  |  |

Devices made from semiconductor materials are small but versatile units that can perform an amazing variety of functions in electronic equipments. In addition, semiconductor devices have many important advantages over other types of electron devices. They are very small and light in weight but they consume very little power.

Like other electronic devices, they have the ability to control almost instantly the movement of charges of electricity. They are used as rectifiers, detectors, amplifiers, oscillators, electronic switches, mixers, and modulators.

The materials most often used in semiconductor devices are germanium (Ge) and silicon (Si). Silicon (Si) and germanium (Ge) are widely used as elemental semiconductors. They have a diamond lattice structure. This structure also belongs to the cubic-crystal family.

More than 90% of the earth's crust is composed of Silica ($SiO_2$) or Silicate, making silicon the second most abundant element on earth. When sand glitters in sunlight, that's silica. Silicon is found in myriad compounds in nature and industry. Most importantly to technology, silicon is the principle platform for semiconductor devices. Silicon is the principal component of most semiconductor devices, especially integrated circuits and microchips. Silicon in the form of a single crystal wafer is the basic building block for the integrated circuit (IC) fabrication.

To keep pace with the growth in IC processing technology, chip size and circuit complexity, silicon crystal and wafers have to be prepared with sustaining increases in diameter and improvements in performation. Silicon used in semiconductor devices manufacturing is currently fabricated into boules that are large enough in diameter to allow the production of 300mm (12inch) wafers. Figure 3-1 shows a piece of silicon wafer. Ultra large scale integration (ULSI) ICs, fabricated with design rules approaching 60nm, have to depend on the availability of highly perfect single crystals, which are prepared exclusively from silicon pulled from the melt by the Czochralski (CZ) technique [2].

Figure 3-1　A piece of silicon wafer

Silicon is widely used in semiconductors because it remains a semiconductor at higher temperatures than the germanium and because its native oxide is easily grown in a furnace and forms a better semiconductor interface than any other material. Its combination of low raw material cost, relatively simple processing, and a useful temperature range make it currently the best compromise among the various competing materials.

Figure 3-2 shows the first transistor, discovered by Bardeen, Brattain and Shockley at Bell Labs in 1947.

This was perhaps the most important electronics event of the 20th century, as it later made possible the integrated circuit and microprocessor. The triangular hunk is a piece of Ge crystal.

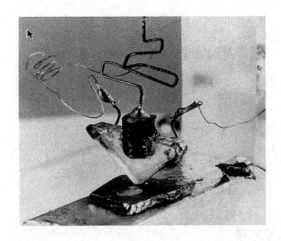

Figure 3-2 The configuration of the first transistor

Germanium (Ge) is an important semiconductor material used in transistors and various other electronic devices too. It was a widely used early semiconductor material but its thermal sensitivity makes it less useful than silicon.

Today, germanium is often alloyed with silicon for use in very-high-speed SiGe devices; IBM is a major producer of such devices.

Gallium Arsenide (GaAs) is also widely used in high-speed devices but so far, it has been difficult to form large-diameter boules of this material, limiting the wafer diameter to sizes significantly smaller than silicon wafers thus making most of GaAs devices significantly more expensive than silicon. Other less common materials are also in use or under investigation. Silicon carbide (SiC) has found some application as the raw material for blue light-emitting diodes (LEDs) and is being investigated for use in semiconductor devices that could withstand very high operating temperatures and environments with the presence of significant levels of ionizing radiation. IMPATT diodes have also been fabricated from SiC.

## New Words

resistivity [ˌriːzisˈtiviti] *n.* 抵抗力，电阻系数
conductivity [ˌkɔndʌkˈtiviti] *n.* 传导性，传导率
periodic [ˌpiəriˈɔdik] *adj.* 周期的，定期的
versatile [ˈvəːsətail] *adj.* 通用的，万能的，多才多艺的，多面手的
instantly [ˈinstəntli] *adv.* 立即地，即刻地
glitter [ɡˈlitə] *vi.* 闪闪发光，闪烁，闪光；*n.* 闪光
myriad [ˈmiriəd] *n.* 无数，无数的人或物；*adj.* 无数的，一万的，种种的
platform [ˈplætfɔːm] *n.* (车站)月台，讲台，讲坛，平台
furnace [ˈfəːnis] *n.* 炉子，熔炉
withstand [wiðˈstænd] *vt.* 抵挡，经受住
radiation [ˌreidiˈeiʃən] *n.* 发散，发光，发热，辐射，放射，放射线，放射物

## Phrases & Expressions

made from　由……制造
belong to　属于
keep pace with　并驾齐驱,保持同步,与……齐步前进
a piece of　一块,一片

## Technical Terms

ULSI　特大规模集成电路
IC　集成电路
IMPATT diode　碰撞雪崩及渡越时间二极管
LED　发光二极管
diamond lattice　晶格
elemental semiconductor　元素半导体
compound semiconductor　化合物半导体
electronic switch　电子开关

## Notes

1. Semiconductor is a material that has a resistivity value between that of a conductor and an insulator. The conductivity of a semiconductor material can be varied under an external electrical field.

半导体是一种导电能力介于导体和绝缘体之间的材料,其导电能力在外部电场的作用下是变化的。

2. Ultra large scale integration (ULSI) ICs, fabricated with design rules approaching 60 nm, have to depend on the availability of highly perfect single crystals, which are prepared exclusively from silicon pulled from the melt by the Czochralski (CZ) technique.

现代的特大规模集成电路,其设计工艺已经要求达到60nm,这是由于单晶硅具有非常良好的性质,而且这些硅片都是用坩埚直拉法形成的。

## Lesson 4　Moore's Law

In April of 1965, Electronics magazine published an article by Intel co-founder Gordon Moore. "Cramming more components onto integrated circuits"— Moore's Law summed up the pace of progress in information technology. In the past 40 years, the computer has become an indispensable tool for the majority of people instead of the mystery of the monster[1]. Information technology goes from the laboratory into the countless ordinary families. Internet links to the world, multi-media audio-visual equipment enrich everyone's lives.

Moore's Law describes a long-term trend in the history of computing hardware. Since the invention of the integrated circuit in 1958, the number of transistors that can be placed inexpensively on an integrated circuit has increased exponentially, doubling approximately every two years. It has continued for almost half a century and was not expected to stop for at least another decade. Almost every measure of the capabilities of digital electronic devices is strongly linked to Moore's Law, such as processing speed, memory capacity, etc[2].

### The origins of Moore's Law

The part of Moore's original 1965 paper that's usually cited in support of this formulation is the following Figure 4-1. This graph does indeed show transistor densities doubling every 12 months, so the formulation above is accurate. However, it doesn't quite do justice to the full scope of the picture that Moore painted in his brief, uncannily prescient paper. This is because Moore's paper dealt with more than just shrinking transistor sizes. Moore was ultimately interested in shrinking transistor costs, and in the effects that cheap, ubiquitous computing power would have on the way we live and work. Figure 4-2 shows the origins of Moore's Law.

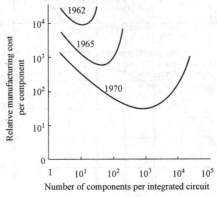

Figure 4-1　Gordon Moore's original graph from 1965

### The effects of Moore's Law

Several measures of digital technology are improving related to Moore's Law, including the size, cost, density and speed of components.

#### 1. Transistors per integrated circuit

The most popular formulation is of the doubling of the number of transistors on integrated circuits every two years. At the end of the 1970s, Moore's Law became known as the limit for the number of transistors on the most complex chips.

Figure 4-3 shows transistors per integrated circuit. Recent trends show that this rate has been maintained into 2007.

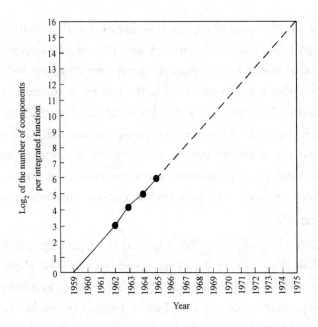

Figure 4-2　The origins of Moore's Law

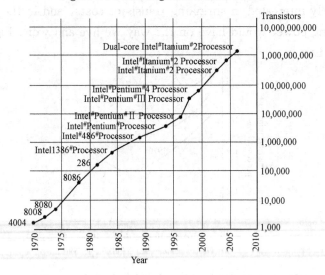

igure 4-3　Transistors per integrated circuit in 1970s—2010s

## 2. Cost per transistor

As the size of transistors has decreased, the cost per transistor has decreased as well. However, the manufacturing cost per unit area has only increased over time, since materials and energy expenditures per unit area have only increased with each successive technology node.

## 3. Computing performance per unit cost

As the size of transistors shrinks, the speed at which they operate increases. Moore's Law is refer to the rapidly continuing advance in computing performance per unit cost, because increase in transistor count is also a rough measure of computer processing performance.

4. Power consumption

The power consumption of computer doubles every 18 months.

### The future of Moore's Law

Computer industry technology "road maps" predict that Moore's Law will continue for several chip generations. Intel has kept that pace for nearly 40 years. Today, they continue to help move the industry forward by delivering:

(1) A worldwide silicon fab network with seven high volume fabs;

(2) The world's first 32nm silicon technology on-target for delivery in 2009;

(3) The world's first 2-billion transistor microprocessor delivered in next-generation Intel® Itanium® processors codenamed Tukwila[3];

(4) Revolutionary technologies on a chip, like hafnium-based high-k metal gate in production today;

(5) Advanced research into trigate transistors and silicon nanotechnology.

## New Words

cram [kræm] v. 填满
mystery ['mistəri] n. 神秘，神秘的事物
monster ['mɔnstə] n. 怪物，妖怪
doubling ['dʌbliŋ] n. 加倍，成双
formulation [ˌfɔːmjuˈleiʃən] n. 用公式表示，明确地表达，做简洁陈述
ubiquitous [juːˈbikwitəs] adj. 到处存在的，(同时)普遍存在的
successive [səkˈsesiv] adj. 继承的，连续的

## Phrases & Expressions

deal with　安排,处理,涉及
relate to　涉及,与……相关的,叙述,和……相处
at the end of　在……结尾,在……末端

## Technical Terms

Moore's Law　摩尔定律
information technology　信息技术
unit area　单元面积
silicon nanotechnology　硅纳米技术
transistor size　晶体管尺寸
integrated circuit　集成电路
memory capacity　存储容量
power consumption　功耗

## Notes

1. In the past 40 years, the computer has become an indispensable tool for the majority of people instead of mystery of the monster.

过去的 40 年里，计算机从神秘的庞然大物发展成为多数人都不可或缺的工具。

2. Almost every measure of the capabilities of digital electronic devices is strongly linked to Moore's Law, such as processing speed, memory capacity, etc.

大多数数字电子元器件的性能与摩尔定律联系非常紧密，如处理速度、存储容量等。

3. Tukwila：英特尔安腾系列的下一代处理器将集成 200 亿个晶体管，由 30MB 板载缓存来支持最多 8 个指令线程，它将是 Intel 首款采用新的 QuickPath（快速路径）互连技术，也就是整合内存控制器的安腾处理器。

## Exercises

1. Keywords

In article, there are some important words which are the soul of this paper. After reading this paper, we can find some words to stand for this article. Now please find out key words of every lesson respectively.

2. Summary

After reading this paper, please write a summary of 200~300 words about Moore's Law.

# 科技英语知识 1：数学符号及数学表达式的读法

| 符号和表达式 | 英语读法 |
|---|---|
| 3/5 | three fifths 或 three over five |
| 25.74 | twenty-five point seven four 或 two five point seven four |
| 5% | five percent |
| $8 \times 10^7$ | eight times the seventh power of ten |
| + | plus 或 positive |
| − | minus 或 negative |
| × | times 或 multiplied by |
| ÷ | over 或 divided by |
| ± | plus or minus |
| = | is equal to 或 equals |
| ≈（及≅） | is approximately equal to 或 approximately equals |
| ≠ | not equal to |
| → | approaches |
| ( ); [ ]; { } | round brackets 或 parentheses; square brackets; braces |
| j | imaginary 或 square root of −1 |
| $f(\ )$ | function $f$ of |
| $x^n$ | the $n$th power of $x$ 或 $x$ to the power $n$ |
| $\sqrt{x}$; $\sqrt[3]{x}$; $\sqrt[n]{x}$ | the square root of $x$; the cube root of $x$; the $n$th root of $x$ |
| $\sin x$; $\arcsin x$ | sine of $x$; arc sine of $x$ |
| $\log_n x$ | log $x$ to the base $n$ |
| $\ln x$ | log $x$ to the base（即 natural logarithms） |
| $\Sigma$; $\Pi$ | the sum of the terms indicated; the product of the terms indicated |
| $x!$ | factorial of $x$ |
| $|x|$ | the absolute value of $x$ |
| $\bar{x}$ | the mean value of $x$ 或 $x$ bar |
| $b'$; $b''$ | $b$ prime; $b$ second prime |
| $b_1$; $b_2$ | $b$ sub one; $b$ sub two |
| $b''_m$ | $b$ second prime sub $m$ |
| $dy/dx$ | the first derivative of $y$ with respect to $x$ |
| $d^n y/dx^n$ | the $n$th derivative of $y$ with respect to $x$ |
| $\partial y/\partial x$ | the partial derivative of $y$ with respect to $x$ |
| $\int_a^b$ | integral between limits $a$ and $b$ |
| $x \to \infty$ | $x$ approaches to infinity |
| $(a+b)$ | bracket $a$ plus $b$ bracket closed |

# Unit 3　　Semiconductor Device

## Lesson 5　Resistors, Capacitors and Inductors

　　This section introduces several of the most common types of electronic component, including resistors, capacitors and inductors. These are often referred to as passive components as they cannot, by themselves, generate voltage or current. In other words, resistors, inductors, and capacitors are passive elements which take energy from the sources and either convert it to another form or store it in an electric or magnetic field. An understanding of the characteristics and application of passive components is an essential prerequisite to understanding the operation of the circuits used in amplifiers, oscillators, filters and power supplies.

### Resistors

　　The amount of current that will flow in a conductor when a given electromotive force (e. m. f.) is applied is inversely proportional to its resistance. Therefore, resistance may be thought of as an opposition to current flow; the higher the resistance the lower the current that will flow (assuming that the applied e. m. f. remains constant). Conventional forms of resistor obey a straight line law when voltage is plotted against current and this allows us to use resistors as a means of converting current into a corresponding voltage drop, and vice versa[1] (note that doubling the applied current will produce double the voltage drop, and so on). Therefore, resistors provide us with a means of controlling the currents and voltages present in electronic circuits. They can also act as loads to simulate the presence of a circuit during testing (a suitably rated resistor can be used to replace a loudspeaker when an audio amplifier is being tested).

　　The resistance of an electrical conductor depends on 4 factors, these being:

　　(1) The length of the conductor. Resistance, $R$, is directly proportional to length, $l$, of a conductor, i. e. $R \propto l$.

　　(2) The cross-sectional area of the conductor. Resistance, $R$, is inversely proportional to cross-sectional area, $a$, of a conductor, i. e. $R \propto 1/a$.

　　(3) The type of material.

　　(4) The temperature of the material.

　　Since $R \propto l$ and $R \propto 1/a$ then $R \propto l/a$. By inserting a constant of proportionality into this relationship the type of material used maybe taken into account. The constant of proportionality is known as the resistivity of the material and is given the symbol $\rho$ (Greek rho). Thus,

$$R = \frac{\rho l}{a}$$

where, $\rho$ is measured in ohm metres ($\Omega$m).

Other practical consideration when selecting resistors for use in a particular application include temperature coefficient. In general, as the temperature of a material increases, most conductors increase in resistance, insulators decrease in resistance, whilst the resistance of some special alloys remains almost constant.

There are also some practical considerations such as noise performance, stability and ambient temperature range.

### Capacitors

Every system of electrical conductors possesses capacitance. A capacitor is a device for storing electric charge. In effect, it is a reservoir into which charge can be deposited and then later extracted. Typical applications include reservoir and smoothing capacitors for use in power supplies, coupling AC signals between the stages of amplifiers, and decoupling supply rails.

The charge or quantity of electricity that can be stored in the electric field between the capacitor plates is proportional to the applied voltage and the capacitance of the capacitor. Thus:

$$Q = CV$$

where $Q$ is the charge (in coulombs), $C$ is the capacitance (in farads), and $V$ is the potential difference (in volts). The unit of capacitance is the farad F.

The main types of the capacitor include: variable air, mica, paper, ceramic, plastic, titanium oxide and electrolytic.

1. Variable air capacitors

These usually consist of two sets of metal (such as aluminum) plates and one fixed the other variable. The set of moving plates rotate on a spindle as shown by the view of Figure 5-1.

2. Mica capacitors

A typical older type construction is shown in Figure 5-2. Usually the whole capacitor is impregnated with wax and placed in a bakelite case. Mica is easily obtained in thin sheets and is a good insulator. However, mica is expensive and is not used in capacitors above about 0.2μF. A modified form of mica capacitor is the silvered mica type.

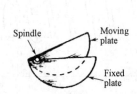

Figure 5-1  Variable air capacitor

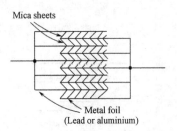

Figure 5-2  The construction of mica capacitors

### 3. Paper capacitors

A typical paper capacitor is shown in Figure 5-3 where the length of the roll corresponds to the capacitance required.

### 4. Ceramic capacitors

These are made in various forms, each type of construction depending on the value of capacitance required. Certain ceramic materials have a very high permittivity and this enables capacitors of high capacitance to be made which are of small physical size with a high working voltage rating. Ceramic capacitors are available in the range 1pF to 0.1μF and may be used in high frequency electronic circuits subject to a wide range of temperatures.

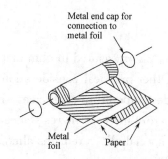

Figure 5-3  A typical paper capacitor

### Inductors

The basic form of an inductor is simply a coil of wire. Factors which affect the inductance of an inductor include:

(1) The number of turns of wire—the more turns the higher the inductance;

(2) The cross-sectional area of the coil of wire—the greater the cross-sectional area the higher the inductance;

(3) The presence of a magnetic core—when the coil is wound on an iron core the same current sets up a more concentrated magnetic field and the inductance is increased;

(4) The way the turns are arranged—a short thick coil of wire has a higher inductance than a long thin one.

The standard electrical circuit diagram symbols for air-cored and iron-cored inductors are shown in Figure 5-4.

An iron-cored inductor is often called a choke since, when used in a.c. circuits, it has a hoking effect, limiting the current flowing through it. Inductance is often undesirable in a circuit. To reduce inductance to a minimum the wire may be bent back on itself, as shown in Figure 5-5, so that the magnetizing effect of one conductor is neutralized by that of the adjacent conductor. The wire may be coiled around an insulator, without increasing the inductance.

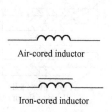

Figure 5-4  The diagram of different inductors

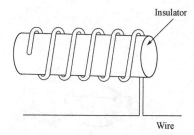

Figure 5-5  The bend wire

## New Words

component [kəm'pəunənt] n. 成分；adj. 组成的，构成的
prerequisite ['priː'rekwizit] n. 先决条件；adj. 首要必备的
proportionality [prəˌpɔːʃə'næliti] n. 比例(性)
essential [i'senʃəl] adj. 必不可少的，绝对必要的
convert [kən'vəːt] vt., vi. (使)转变，(使)转化
wax [wæks] n. 蜡
mica ['maikə] n. 云母
ceramic [si'ræmik] adj. 陶器的；n. 陶瓷制品
coil [kɔil] v. 盘绕，卷

## Phrases & Expressions

in other words 换句话说
in general 通常，大体上，一般而言
take into account 重视，考虑

## Technical Terms

amplifier ['æmpliˌfaiə(r)] n. 放大器
oscillator ['ɔsileitə] n. 振荡器
filter ['filtə] n. 滤波器
permittivity [ˌpəmi'tiviti] n. [电]介电常数，电容率
loudspeaker ['laud'spiːkə] n. 扩音器
farad ['færəd] n. 法拉(电容单位)
choke [tʃəuk] n. 扼流圈
passive component 无源元件
titanium oxide 氧化钛陶瓷
magnetizing effect 电磁效应
electromotive force 电动势
ambient temperature range 周围温度范围
iron-cored inductor 铁芯电感

## Notes

1. Conventional forms of resistor obey a straight line law when voltage is plotted against current and this allows us to use resistors as a means of converting current into a corresponding voltage drop, and vice versa.

电阻的电流-电压特性曲线通常遵从直线规律,所以我们可以通过电阻把电流转换成为相应的压降。

在这个长句中,由 and 连接两个句子构成。前一句是一个 when 引导时间状语从句,主句中 conventional forms of resistor 是主语,obey 是谓语,a straight line law 是宾语,在从句中 voltage 是主语,is plotted against 是谓语,current 是宾语。后句中,this 是主语,指代前面的整个句子,allows us to use 是谓语,as a means of ... 做状语修饰宾语 resistors。

## Lesson 6  Diode

A diode is an electrical device allowing current to move through it in one direction with far greater ease than in the other. The most common type of diode in modern circuit design is the semiconductor diode, although other diode technologies exist. A semiconductor diode is a device having a p-n junction mounted in a container, suitable for conducting and dissipating the heat generated in operation, and having connecting leads. Two circuit diagram symbols for semiconductor diodes are in common use and are as shown in Figure 6-1.

Figure 6-1  Two circuit diagram symbols of semiconductor diodes

Unlike a Schottky diode (a majority carrier device), a p-n junction diode is known as a minority carrier device since the current conduction is controlled by the diffusion of minority carriers (electrons in the p region and holes in the n region) in a p-n junction diode. A p-n junction diode can be fabricated by doping the semiconductor material with opposite doping impurities (acceptor or donor impurities) to form the p and n regions of the diode. If a p-n junction is formed on the same semiconductor it is referred to as a p-n homojunction diode. On the other hand, if a p-n junction is formed using two semiconductor materials of different band gaps and with opposite doping impurities, then it is referred to as a p-n heterojunction diode.

The applications of diode are based on certain properties of the junction:

(1) The injection of electron-hole pairs to generate light via recombination (LEDs and LASERs);

(2) The separation of electron-hole pairs at the junction to constitute a current source (solar cell);

(3) The temperature dependence of the *I-V* characteristic (a temperature sensor);

(4) The non-linear nature of the *I-V* characteristic (frequency multipliers and mixers);

(5) The device as a switch (rectifiers, inverters, power supplies etc.).

Let's take a look at the simple battery-diode-lamp circuit. When the polarity of the battery is such that electrons are allowed to flow through the diode, the diode is said to be forward-biased. Conversely, when the battery is "backward" and the diode blocks current, the diode is said to be reverse-biased. A diode may be thought of as a kind of switch: "close" when forward-biased and "open" when reverse-biased. The essential difference between forward-biased and reverse-biased is the polarity of the voltage dropped across the diode. When the diode is forward-biased and conducting current, there is a small voltage dropped across it, leaving most of the battery voltage dropped across the lamp. In other words, under forward-biased conditions, the current increases exponentially with applied voltage. When the battery's polarity is reversed and the diode becomes reverse-biased, it drops all of the battery's voltage and leaves none for the lamp.

In fact, this forward-biased voltage drop exhibited by the diode is due to the action of the depletion region formed by the p-n junction under the influence of an applied voltage. When there is no voltage applied across a semiconductor diode, a thin depletion region exists around the region of the p-n junction, preventing current through it. The depletion region is for the most part devoid of available charge carriers and so acts as an insulator.

If a reverse-biased voltage is applied across the p-n junction, this depletion region expands (shown as Figure 6-2), further resisting any current through it.

Conversely, if a forward-biased voltage is applied across the p-n junction, the depletion region will collapse and become thinner(shown as Figure 6-3), so that the diode becomes less resistive to current through it[1]. In order for a sustained current to go through the diode, though, the depletion region must be fully collapsed by the applied voltage. This takes a certain minimum voltage to accomplish, called the threshold voltage.

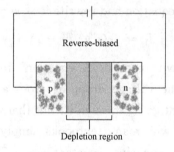

Figure 6-2  The depletion region of reverse-biased

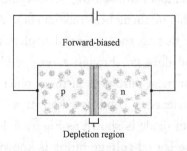

Figure 6-3  The depletion region of forward-biased

For silicon diodes, the typical threshold voltage is 0.7V, nominally. For germanium diodes, the threshold voltage is only 0.3V. The chemical constituency of the p-n junction comprising the diode accounts for its nominal threshold voltage figure, which is why silicon and germanium diodes have such different threshold voltages[2]. Forward voltage drop remains approximately equal for a wide range of diode currents, meaning that diode voltage drop not like that of a resistor or even a normal (closed) switch. For most purposes of circuit analysis, it may be assumed that the voltage drop across a conducting diode remains constant at the nominal figure and is not related to the amount of current going through it.

In actuality, things are more complex than this. There is an equation describing the exact current through a diode, given the voltage dropped across the junction, the temperature of the junction, and several physical constants. It is commonly known as the diode equation:

$$I_D = I_S[e^{qV_D/(NkT)} - 1] \qquad (6\text{-}1)$$

where, $I_D$ = Diode current in amps, $I_S$ = Saturation current in amps (typically $1 \times 10^{-12}$ amps), e = Euler's constant ($\sim 2.718281828$), $q$ = charge of electron ($1.6 \times 10^{-19}$ coulombs), $V_D$ = Voltage applied across diode in volts, $N$ = "Nonideality" or "emission" coefficient (typically between 1 and 2), $k$ = Boltzmann's constant ($1.38 \times 10^{-23}$), $T$ = Junction temperature in degrees Kelvin.

The equation $kT/q$ describes the voltage produced within the p-n junction due to the action of temperature, and is called the thermal voltage, or $V_t$ of the junction. At room temperature, this is about 26 millivolts. Knowing this, and assuming a "nonidealiy" coefficient of 1, we may simplify the diode equation and rewrite it as such:

$$I_D = I_S(e^{V_D/0.026} - 1) \qquad (6\text{-}2)$$

You need not be familiar with the "diode equation" in order to analyze simple diode circuits. Just understand that the voltage dropped across a current-conducting diode does change with the amount of current going through it, but that this change is fairly small over a wide range of currents. This is why many textbooks simply say the voltage drop across a conductor, semiconductor diode remains constant at 0.7V for silicon and 0.3V for germanium. Also, since temperature is a factor in the diode equation, a forward-biased p-n junction may also be used as a temperature-sensing device, and thus can only be understood if one has a conceptual grasp on this mathematical relationship. A reverse-biased diode prevents current from going through it, due to the expanded depletion region. In actuality, a very small amount of current can and does go through a reverse-biased diode, called the leakage current, but it can be ignored for most purposes. If the applied reverse-biased voltage becomes too great, the diode will experience a condition known as breakdown, which is usually destructive. The I-V character of diode is shown as Figure 6-4.

A diode's maximum reverse-biased voltage rating is known as the Peak Inverse Voltage, or PIV, and may be obtained from the manufacturer. Like forward voltage, the PIV rating of a diode varies with temperature, except that PIV increases with increased temperature and decreases as the diode becomes cooler exactly opposite that of forward voltage. Typically, the PIV rating of a generic "rectifier" diode is at least 50V at room temperature.

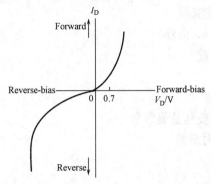

Figure 6-4　The *I-V* character of diode

## New Words

polarity [pəuˈlæriti] *n.* 极性
coefficient [ˌkəuiˈfiʃənt] *n.* [数]系数
impurity [imˈpjuəriti] *n.* 杂质
exhibit [igˈzibit] *vt.* 展出，陈列
collapse [kəˈlæps] *n.* 倒塌，崩溃
comprise [kəmˈpraiz] *v.* 包含，由……组成
destructive [disˈtrʌktiv] *adj.* 破坏(性)的
manufacturer [ˌmænjuˈfæktʃərə] *n.* 制造商，制造厂
injection [inˈdʒekʃən] *n.* 注射，注入
mathematical [ˌmæθəˈmætikl] *adj.* 数学的，数学上的

## Phrases & Expressions

be known as　被认为是
base on　基于
take a look　注视，看一看，瞧一瞧
act as　担当，充当，担任
in order to　为了
due to　由于，应归于

## Technical Terms

diode [ˈdaiəud] *n.* 二极管

semiconductor [ˈsemikənˈdʌktə] *n.* 半导体
laser [ˈleizə] *n.* 激光器
rectifier [ˈrektifaiə] *n.* 整流器
breakdown [ˈbreikdaun] *n.* 击穿
forward-biased    正向偏置
voltage drop    压降
depletion region    耗尽层
Boltzmann's constant    玻耳兹曼常数
emission coefficient    发射系数
thermal voltage    热电压
room temperature    室温
leakage current    漏电流
peak inverse voltage    峰值反向电压
Schottky diode    肖特基二极管
minority carrier    少数载流子
heterojunction diode    异质结二极管
homojunction diode    同质结二极管
threshold voltage    阈值电压

# Notes

1. Conversely, if a forward-biased voltage is applied across the p-n junction, the depletion region will collapse and become thinner(shown as Figure 6-3), so that the diode becomes less resistive to current through it.

相反，如果在 p-n 结上施加一个正向偏置电压，耗尽层就会塌陷，变得更薄（如图 6-3 所示），因此二极管对电流的阻碍就减小了。

这个句子的框架结构是... the depletion region will collapse and become thinner ,,, If... 引导条件状语从句中 voltage 是主语，is applied across 是谓语，the p-n junction 为宾语；so that 用来引导结果状语从句，意思为"因此，结果"。

2. The chemical constituency of the p-n junction comprising the diode accounts for its nominal threshold figure, which is why silicon and germanium diodes have such different threshold voltages.

构成二极管 p-n 结的化学成分决定了它的阈值电压，也就表明了为什么硅二极管和锗二极管的阈值电压不同。

这个句子的框架结构是 The chemical constituency... accounts for its nominal forward voltage figure. 关系代词 which 引导非限制性定语从句，其先行词是前面一整句。在非限制性定语从句中，why 引导表语从句，强调结果。

# Lesson 7  Transistor

A transistor is a linear semiconductor device that controls current with the application of a lower-power electrical signal. Transistors may be roughly grouped into two major divisions: bipolar and field-effect. The bipolar junction transistor was the first three-terminal device in solid state electronics and continues to be a device of choice for many digital and microwave applications. For a decade after its invention, the bipolar device remained the only three-terminal device in commercial applications. The bipolar transistor utilize a small current to control a large current. The field-effect transistor is a device which utilizing a small voltage to control current. All field-effect transistors are unipolar rather than bipolar devices. These simple devices are majority carrier devices which are relatively simple to fabricate and are extremely versatile. FETs are now made from a wide variety of materials (Si, SiGe, GaAs, InGaAs, GaN, SiC, etc.).

### Bipolar junction transistor

The invention of the bipolar transistor in 1948 ushered in a revolution in electronics. The bipolar junction transistor consists of three regions of semiconductor material. One type is called a p-n-p transistor, in which two regions of p-type material sandwich a very thin layer of n-type material. A second type is called an n-p-n transistor, in which two regions of n-type material sandwich a very thin layer of p-type material. Both of these types of transistors consist of two p-n junctions placed very close to one another in a back-to-back arrangement on a single piece of semiconductor material. Diagrams of these two types of transistors are shown in Figure 7-1.

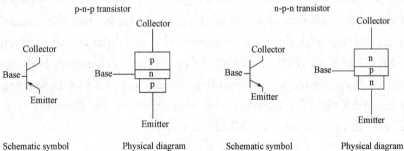

Figure 7-1  The schematic symbols and physical diagrams of two transistor types

The two p-type material regions of the p-n-p transistor are called the emitter and collector and the n-type material is called the base. Similarly, the two n-type material regions of the n-p-n transistor are called the emitter and collector and the p-type material region is called the base.

The only functional difference between a p-n-p transistor and an n-p-n transistor is the proper biasing (polarity) of the junctions when operating. For any given state of operation, the current directions and voltage polarities for each type of transistor are exactly opposite each other. Bipolar transistors work as current-controlled current regulators. In other words, they restrict the amount of current that can go through them according to a smaller, controlling current (shown as Figure 7-2). The main current that is controlled goes from collector to emitter, or from emitter to collector, depending on the type of transistor it is (p-n-p or n-p-n, respectively). According to the confusing standards of semiconductor symbol, the arrow always points against the direction of electron flow.

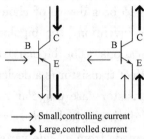

→ Small, controlling current
→ Large, controlled current

Figure 7-2  Different types of current-controlling

### Field-effect transistor

The basic concept behind the FET is quite simple. A FET has three terminals, as shown as Figure 7-3, a source, a drain and a gate. The semiconductor region between source and drain is called a channel and its conductivity is controlled by the potential of the gate terminal. The source and drain contacts are ohmic contacts. It is important to isolate the gate from the channel so that no current flows into the gate. The gate isolation is done in a variety of ways, leading to a number of different devices. There exist two types of FET, the junction-gate device (JUGFET or JFET), in which the gate is separated from the channel by a (normally reverse biased) p-n junction, and the insulated-gate FET (IGFET) in which a thin layer, usually of silicon oxide, provides isolation of the gate electrode from the channel[1]. Because of its structure, this latter device is also termed the metal-oxide-semiconductor transistor (MOSFET or MOST). Both JFETs and MOSFETs exhibit an extremely high input resistance which accounts for their popularity in amplifier circuits and the low power requirements of logic circuitry using MOSFETs, together with very small device dimensions, make possible the very large-scale integrated circuits (VLSI) of today.

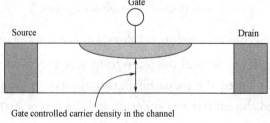

Figure 7-3  The symbol of FET

### Transistor characteristics

1) Transistor connections

There are three ways of connecting a transistor, depending on the use to which it is being put. The ways are classified by the electrode that is common to both the input and the output. They are called: (a) common-base configuration, shown in Figure 7-4(a); (b) common-emitter configuration, shown in Figure 7-4(b); (c) common-collector configuration, shown in Figure 7-4(c).

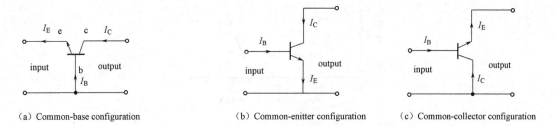

(a) Common-base configuration  (b) Common-enitter configuration  (c) Common-collector configuration

Figure 7-4　Different configurations of transistor

These configurations are for an n-p-n transistor. The current flows shown are all reversed for a p-n-p transistor.

2) Transistor characteristics

The effect of changing one or more of the various voltages and currents associated with a transistor circuit can be shown graphically and these graphs are called the characteristics of the transistor[2]. As there are five variables (collector, base and emitter currents and voltages across the collector and base and emitter and base) and also three configurations, many characteristics are possible. Some of the possible characteristics are given below.

(1) Common-base configuration

① Input characteristic

With reference to Figure 7-4 (a), the input to a common-base transistor is the emitter current, $I_E$, and can be varied by altering the base emitter voltage $V_{EB}$. The base-emitter junction is essentially a forward biased junction diode, so as $V_{EB}$ is varied, the current flowing is similar to that for a junction diode, as shown in Figure 7-5 for a silicon transistor.

Figure 7-5 is called the input characteristic for a n-p-n transistor having common-base configuration. The variation of the collector-base voltage $V_{CB}$ has little effect on the characteristic. A similar characteristic can be obtained for a p-n-p transistor, these having reversed polarities.

② Output characteristics

The value of the collector current $I_C$ is very largely determined by the emitter current, $I_E$. For a given value of $I_E$ the collector-base voltage, $V_{CB}$, can be varied and has little effect on the value of $I_C$. If $V_{CB}$ is made slightly negative, the collector no longer attracts the majority carriers leaving the emitter and $I_C$ falls rapidly to zero. A family of curves for various values of $I_E$ are possible and some of these are shown in Figure 7-6.

Figure 7-6 is called the output characteristics for an n-p-n transistor having common-base configuration. Similar characteristics can be obtained for a p-n-p transistor, these having reversed polarities.

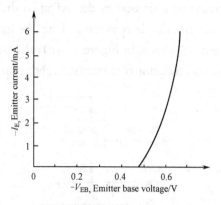

Figure 7-5   The input characteristic for an n-p-n transistor having common-base

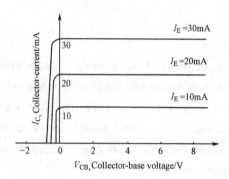

Figure 7-6   The output characteristics for an n-p-n transistor having common-base configuration

(2) Common-emitter configuration

① Input characteristic

In a common-emitter configuration (see Figure 7-4(b)), the base current is now the input current. As $V_{EB}$ is varied, the characteristic obtained is similar in shape to the input characteristic for a common-base configuration shown in Figure 7-5, but the values of current are far less. As long as the junctions are biased as described, the three currents $I_E$, $I_C$ and $I_B$ keep the ratio $1:\alpha:(1-\alpha)$, whichever configuration is adopted. Thus the base current changes are much smaller than the corresponding emitter current changes and the input characteristic for a n-p-n transistor is as shown in Figure 7-7. A similar characteristic can be obtained for a p-n-p transistor, these having reversed polarities.

② Output characteristics

A family of curves can be obtained, depending on the value of base current $I_B$ and some of these for a n-p-n transistor are shown in Figure 7-8. A similar set of characteristics can be obtained for a p-n-p transistor, these having reversed polarities. These characteristics differ from the common base output characteristics in two ways.

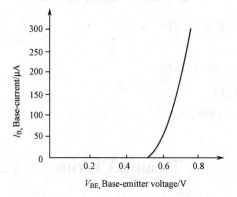

Figure 7-7　The input characteristic of an n-p-n transistor

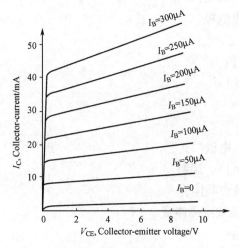

Figure 7-8　The output characteristic of an n-p-n transistor

## New Words

restrict [rist'rikt] *vt.* 限制，约束，限定
symbology [sim'bɔlədʒi] *n.* 象征学，象征的使用
isolation [ˌaisəu'leiʃən] *n.* 隔绝，隔离，绝缘
configuration [kən'figjur'eiʃən] *n.* 构造，形状，外貌
confuse [kənˌfjuːz] *vt.* 使困惑，把……弄糊涂
attract [ə'trækt] *vt.* 吸引
slight [slait] *adj.* 微小的，轻微的
curve [kəːv] *n.* 曲线；*vi.* 呈曲线
utilize [j'uːtilaiz] *vt.* 利用，使用

## Phrases & Expressions

according to    根据……所说；按照……
because of    因为
with reference to    关于，有关，根据
be similar to    与……相似
as long as    只要，在……的时候

## Technical Terms

transistor [træn'zistə] *n.* 晶体管
sandwich ['sænwidʒ, -tʃ] *n.* 三明治
collector [kə'lektə] *n.* 集电极
base [beis] *n.* 基极
emitter [i'mitə] *n.* 发射极
source [sɔːs] *n.* 源极
drain [drein] *n.* 漏极
gate [geit] *n.* 栅极
electrode [i'lektrəud] *n.* 电极
modulation [ˌmɔdju'leiʃən] *n.* 调制
bipolar transistor    双极型晶体管
field-effect transistor    场效应晶体管
ohmic contact    欧姆接触
insulated-gate FET    绝缘栅场效应晶体管
metal-oxide-semiconductor transistor (MOSFET or MOST) 金属氧化物半导体晶体管
very large-scale integrated circuits (VLSI)    超大规模集成电路
common-base    共基极
common-emitter    共发射极
common-collector    共集电极

## Notes

1. There exist two types of FET, the junction-gate device (JUGFET or JFET), in which the gate is separated from the channel by a (normally reverse biased) p-n junction, and the insulated-gate FET (IGFET) in which a thin layer, usually of silicon oxide, provides isolation of the gate electrode from the channel.

有两种类型的场效应晶体管，一种是结-栅型器件（结栅场效应晶体管，或称 JFET），通过一个 p-n 结把栅极与沟道隔离出来（通常是在反向偏置的情况下），而绝缘栅型（即 IGFET）的结

构中有一个小薄层,通常使用二氧化硅,使栅极与沟道隔离。

这个句子的框架结构是 There exist two types of FET, the junction-gate device (JUGFET or JFET),... and the insulated-gate FET (IGFET)...。本句中包括两个由关系代词 in which 引导的非限制定语从句,分别修饰 the junction-gate device (JUGFET or JFET) 和 the insulated-gate FET (IGFET)。

2. The effect of changing one or more of the various voltages and currents associated with a transistor circuit can be shown graphically and these graphs are called the characteristics of the transistor.

晶体管电路中改变一个或多个电压或电流的效果可以用图形的形式表达出来,这些图形就是所谓的晶体管特性曲线。

这个句子的框架结构是 The effect ... can be shown ... and these graphs are called the characteristics of the transistor. "... of changing one or more of the various voltages and currents associated with a transistor circuit..." 修饰主语 effect。

## Exercises

1. Keywords

In the article, there are some important words which are the soul of this paper. After reading this paper, we can find some words to stand for this article. Now please find out key words of this paper.

2. Summary

A transistor is a linear semiconductor device that controls current with the application of a lower-power electrical signal. Do you have some idea with this article? Please write a summary about this paper with less 200 words.

# 科技英语知识 2：形容词的翻译

形容词在英文句中一般做表语或修饰名词的定语，因而在翻译时要把形容词与其所修饰的客体统一考虑，对二者的割裂会导致牛头不对马嘴的错误。

例如，heavy 的基本词义是"重"，但在翻译时结合修饰的对象则译法各异：heavy current 强电流；heavy crop 大丰收；heavy traffic 交通拥挤；heavy industry 重工业。

英语词义对上下文的依赖性是很大的，一个孤立的词，其词义通常是不确定的，但当词处于特定的联立关系中时，其含义就受到相关词的制约而明朗和稳定了。因此，根据词的联立关系确定词义是词义辨析的重要且最为可行的手段。

例如，universal 有"宇宙的"、"世界的"、"普遍的"、"一般的"、"通用的"等含义。具体用法举例如下：

- Universal meter      万用表
- Universal motor      交直流两用电动机
- Universal agent      全权代理人
- Universal peace      世界和平
- Universal truth       普遍真理
- Universal valve      万向阀
- Universal constant     通用常数
- Universal use       普遍应用
- Universal class      全类
- Universal rules      一般法则
- Universal travel      环球旅行
- Universal dividing head   多用分度表头

一般的常用专业词组搭配都可以在英汉技术字典中查到，所以对含义不清楚的词组应借助字典来理解。"闭门造车"式的翻译只能导致主观臆断的错误。对形容词的翻译还可采取词性转换的技巧，使对原文的表达更通顺和易于理解。

1. 当形容词在句中做表语时，某些情况下可根据汉语习惯，将其和系词合译成动词形式。

例：Internet is different from Intranet in many aspects though their spelling is alike.

虽然拼写很像，Internet 仍在很多方面不同于 Intranet。

句中形容词 different 和 alike 分别译为"不同"和"相像。"

例：If low-cost power becomes available from nuclear power plants, the electricity crisis would be solved.

如能从核电站获得低成本电力，电力紧张问题就会解决。

句中 becomes available 译为"获得"。

2. 某些表示事物特征的形容词，可在其后加上"性"、"度"等转译为名词。

例：IPC is more reliable than common computer.

工控机的可靠性比普通计算机高。

例：Experiment indicates that the new chip is about 1.5 times as integrative as that of the old ones.

试验表明新型芯片的集成度是旧型号的1.5倍。

3. 有时形容词也可根据需要译成副词。

例：The same principles of low internet resistance also apply to millimeters.

低内阻原理也同样适用于毫安表。

句中same译为"同样地"。

词性转换技巧同样适用于其他词类，拘泥于原文的词性而进行完全对应的翻译往往会带来极大的障碍，且使得译文生涩而难于理解。翻译应从整体着眼，采取适当的词性转换，运用句子成分转换或句型转换的翻译技巧，可使译文既忠实于原意又通顺可读。

# Unit 4　Operational Amplifier

## Lesson 8　Performance Parameters

The operational amplifier (op-amp), as the basic component is perhaps the most important element in analog circuits and deserves the most attention from the circuit designers. Op-amps are among the most widely used electronic devices today, being used in a vast series of consumer, industrial, and scientific devices. An operational amplifier, which is often called an op-amp, is a DC-coupled high-gain electronic voltage amplifier with differential inputs and, usually, a single output.

The circuit symbol for an op-amp is shown in Figure 8-1, where:

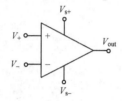

Figure 8-1　Circuit diagram symbol for an op-amp

$V_+$: non-inverting input, $V_-$: inverting input;

$V_{out}$: output;

$V_{s+}$: positive power supply, $V_{s-}$: negative power supply.

The designer of an op-amp must have a clear understanding of what Op-Amp parameters mean and their impact on circuit design. In addition to gain and bandwidth, such parameters as Input offset voltage, slew rate, CMRR etc, may be important. These parameters are the criterions of the amplifier.

### Finite gain

Open-loop gain is defined as the amplification from input to output without any feedback applied[1]. For mathematical calculations, the ideal open-loop gain is infinite; however, it is finite in real operational amplifiers. Typical devices exhibit open-loop DC gain ranging from 100,000 to over 1 million. So long as the loop gain (the product of open-loop and feedback gains) is very large, the circuit gain will be determined entirely by the amount of negative feedback (it will be independent of open-loop gain). In cases where closed-loop gain must be very high, the feedback gain will be very low, and the low feedback gain causes low loop gain; in these cases, the operational amplifier will cease to behave ideally.

### Bandwidth

Finite bandwidth-all amplifiers have a finite bandwidth. This creates several problems for op-amps. First, associated with the bandwidth limitation is a phase difference between the input signal and the amplifier output that can lead to oscillation in some feedback circuits. The internal frequency compensation used in some op-amps to increase the gain or phase margin intentionally reduces the bandwidth even further to maintain output stability when using a wide variety of feedback networks. The high-frequency behavior of op-amps plays a critical role in many applications. Bandwidth is usually defined as the unity-gain frequency. The 3-dB frequency may also be specified to allow easier prediction of the closed frequency response. Figure 8-2 shows the frequency response of magnitude of $A_u$ for an op-amp.

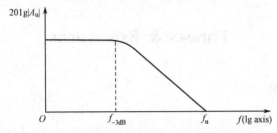

Figure 8-2  The frequency response of magnitude of $A_u$ for an op-amp

### Input offset voltage

All op-amps require a small voltage between their inverting and noninverting inputs to balance mismatches due to unavoidable process variations. The input offset voltage is seen to be typically 1mV, but can go up to as high as 6mV. The output offset voltage is then computed based on the circuit used. If the worst condition is of interest, the maximum value should be used. Typical values are those more commonly expected when using the op-amp.

### Slew rate

Another important factor that contributes to the op-amp's performance is the slew rate (SR). In practice, the slew rate is often referred to as the nonlinear settling rate, and is defined as the maximum rate at which the op-amp's output voltage changes in the presence of a large input signal[2]. It can be expressed as

$$\text{SR} = \left. \frac{dV_{out}}{dt} \right|_{max}$$

### Common-mode rejection ratio (CMRR)

The common-mode rejection ratio is an important figure of merit for operational amplifiers. It is desired to be as large as possible - ideally infinite, infinite The CMRR can be defined as

$$\text{CMRR} = \frac{A_{VD}}{A_{VC}}$$

$A_{VD}$: Gain in differential mode; $A_{VC}$: Gain in common mode

Ideal op-amp CMRR is infinite and real op-amp CMRR is 60~80dB.

## New Words

oscillation [ˌɔsiˈleiʃən] n. 摆动，振动
stability [stəˈbiləti] n. 稳定性
unavoidable [ˈʌnəˈvɔidəbl] adj. 不能避免的，不可避免的，不能取消的
distortion [disˈtɔːʃən] n. 扭曲，变形，曲解，失真
settling [ˈsetliŋ] n. 沉淀物
merit [ˈmerit] n. 优点，价值；vt. 有益于
desire [diˈzaiə] vt. 想望，期望，希望，请求；n. 愿望，心愿，要求

## Phrases & Expressions

so long as　只要
associate with　联合

## Technical Terms

negative feedback　负反馈
unity-gain　单位增益
closed-loop gain　闭环增益
common-mode rejection ratio(CMRR)　共模抑制比
open-loop gain　开环增益
unity-gain frequency　单位增益频率
input offset voltage　输入失调电压
slew rate　压摆率

## Notes

1. Open-loop gain is defined as the amplification from input to output without any feedback applied.

开环增益被定义为放大器没有反馈通路时输入到输出的增益。

2. In practice, the slew rate is often referred to as the nonlinear settling rate, and is defined as the maximum rate at which the op-amp's output voltage changes in the presence of a large input signal.

实际上，压摆率通常被看作是非线性率，定义为当输入大信号时，放大器电路输出电压最大的变化速率。

# Lesson 9   Operational Amplifiers

The block diagram of a general op-amp is shown in Figure 9-1. The block describes op-amp's important parts. In general, it contains input stage, output stage and middle stage. The first stage will be the differential amplifier. This stage can magnify the differential signal. It serves as an excellent input stage. The middle stage will provide high gain . The output stage is to drive an external load without deteriorating the performance of the high-gain amplifier [1]. It is also necessary to provide a stabilized bias for each of the previous stages; biasing stage could be distributed to provide the bias currents to the other stages.

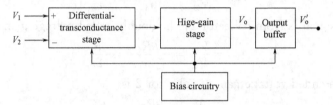

Figure 9-1   The block diagram of an op-amp

### Differential amplifier circuit

The differential amplifier circuit is an extremely popular connection used in IC units. This connection can be described by considering the basic differential amplifier shown in Figure 9-2. Notice that the circuit has two separate inputs and two separate outputs, and that the emitters are connected together. Whereas most differential amplifier circuits use two separate voltage supplies, the circuit can also operate using a singe supply.

### AC operation of circuit

An ac connection of a differential amplifier is shown in Figure 9-3. Seperate input signals are applied as $V_{i1}$ and $V_{i2}$ with separate outputs resulting as $V_{o1}$ and $V_{o2}$.

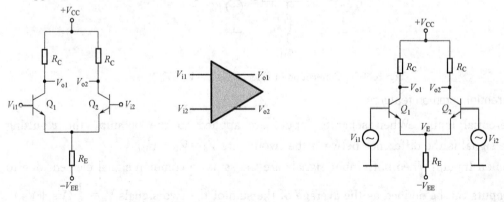

Figure 9-2   Basic differential amplifier circuit         Figure 9-3   AC connection of differential amplifier

To carry out AC analysis, we redraw the circuit in Figure 9-4.

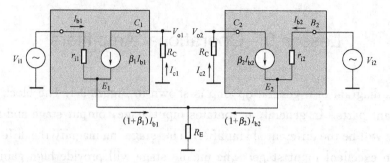

Figure 9-4   AC equivalent of differential amplifier circuit

Double-Ended AC Voltage Gain magnitude from both collectors is

$$A_d = \frac{V_o}{V_{id}} = -\frac{\beta R_C}{r_i} = -\frac{R_C}{r_e}$$

If the output signal is from collector 1, the double-ended voltage gain at collector1 is

$$A_{V1} = \frac{V_{o1}}{V_{id}} = -\frac{\beta R_C}{2r_i} = -\frac{R_C}{2r_e}$$

For the double-ended voltage gain at collector 2 is

$$A_{V2} = \frac{V_{o1}}{V_{id}} = \frac{\beta R_C}{2r_i} = -\frac{R_C}{2r_e}$$

Single-Ended AC Voltage Gain is to calculate $\frac{V_o}{V_i}$, applying signal to one input with the other connected to ground, as shown in Figure 9-5.

$$A_{V1} = \frac{V_{o1}}{V_{i1}} = \frac{\beta R_C}{2r_i} = -\frac{R_C}{2r_e}$$

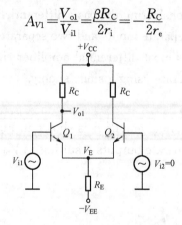

Figure 9-5   Connection of calculate $A_{V1} = V_{o1}/V_{i1}$

**Differential common features**

Differential Input: When separate inputs are applied to the op-amp, the resulting difference signal is the difference between the two inputs $V_d = V_{i1} - V_{i2}$.

Common Inputs: When both input signals are the same, a common signal element due to the two inputs can be defined as the average of the sum of the two signals $V_c = \frac{1}{2}(V_{i1} + V_{i2})$.

### Output voltage

Since any signals applied to an op-amp in general have both in-phase and out-of-phase components, the resulting output can be expressed as $V_o = A_d V_d + A_c V_c$.

Complementary metal-oxide semiconductor (CMOS) technology has been the mainstay in mixed-signal implementations because it provides density and power savings on the digital side, and a good mix of components for analog design. Because of its widespread use. The following context shall be provided in amplifier design examples based CMOS amplifier.

### Differential input stage

#### 1. Traditional differential input stage

Figure 9-6 shows a traditional differential input stage. It uses a current-mirror load and n-channel MOSFETs $M_1$ and $M_2$ to form a differential amplifier. $M_1$ and $M_2$ are biased with a current sink connected to the sources of $M_1$ and $M_2$.

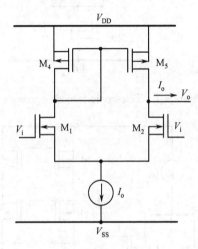

Figure 9-6  Configuration of traditional differential input stage

The traditional differential input stage has an advantage of simple structure, large bandwidth, but it has low gain. In general, the open-loop gain is only 60dB. In order to achieve a high open-loop gain, we often adopt the cascode configurations. These configurations contain telescopic cascode configuration.

#### 2. Telescopic cascode input stage

In order to achieve high gain, the differential cascade topologies of Figure 9-7 can be used. Figure 9-7 shows the configuration of telescopic cascode, This structure pays the cost of output swing and additional poles. Another Drawback of telescopic cascodes is the difficulty in shorting their inputs and outputs to implement a unity-gain buffer [2].

#### 3. Folded-cascode input stage

In order to change the small output swing of telescopic op-amp and difficulty in shorting the input and output, we can choose the folded-cascode op-amp. The folded-cascode op-amp with cascode PMOS loads shown in Figure 9-8. The folded-cascode op-amp circumvents the issue of inherent voltage drop.

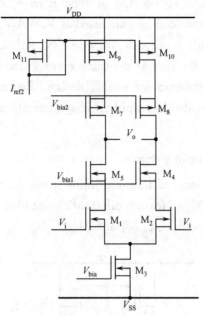

Figure 9-7  Telescopic cascode op-amp

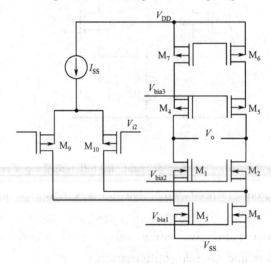

Figure 9-8  Folded-cascade op-amp

### Output stage

The primary objective of output stage is to convert current any drive signals into an output load, there are several approaches to implement. It includes the class A amplifier, source followers, the push-pull amplifier.

### 1. Class A amplifier

Figure 9-9 shows Class A amplifier. The load of this inverter consists of a resistance $R$ and a capacitance $C$. The maximum efficiency of the Class A output stage is 25%.

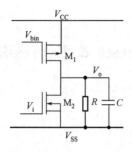

Figure 9-9　Class amplifier

### 2. Source followers

Source followers has both large current gain and low output resistance. Figure 9-10 shows the configuration. Unfortunately, since the source is the output node, the MOS device becomes dependent on the body effect. The distortion of the source follower will be better than the Class A amplifier.

### 3. Push-pull output amplifier

Figure 9-11 shows the push-pull amplifier. An advantage of this circuit is the currents are actively sourced and sunk. A disadvantage is that the static current very high.

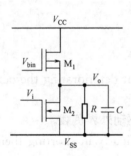
Figure 9-10　Source followers

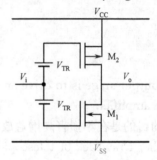
Figure 9-11　Push-pull amplifier

## New Words

differential [ˌdifəˈrenʃəl] *adj.* 微分的；*n.* 微分
magnify [ˈmægnifai] *vt.* 放大，扩大，赞美，夸大，夸张；*vi.* 有放大能力
excellent [ˈeksələnt] *adj.* 卓越的，极好的
extremely [iksˈtriːmli] *adv.* 极端地，非常地
widespread [ˈwaidspred,-ˈspred] *adj.* 分布广泛的，普遍的
buffer [ˈbʌfə] *n.* 缓冲器
pole [pəul] *n.* 棒，柱，杆，竿，极，磁极，电极；*vt.* 用竿支撑，用棒推；*vi.* 撑篙
efficiency [iˈfiʃənsi] *n.* 效率，功效

## Phrases & Expressions

carry out    完成,实现,贯彻,执行
by reason of    由于,因为
consist of    由……组成

## Technical Terms

telescopic cascode    套筒共源共栅
folded-cascode    折叠共源共栅
source follower    源极跟随器
differential amplifier    差分放大器
current sink    电流沉
unity-gain buffer    单位增益缓冲器
push-pull amplifier    推挽放大器

## Notes

1. The output stage is to drive an external load without deteriorating the performance of the high-gain amplifier.

输出级的目的是在不损害高增益放大器性能的条件下驱动外界负载。

2. Another Drawback of telescopic cascodes is the difficulty in shorting their inputs and outputs to implement a unity-gain buffer.

套筒共源共栅结构的另一个缺点是很难以输入和输出短路的方式实现单位增益缓冲器.

# Lesson 10  Op-amp Applications

The operational amplifier was originally given to an amplifier that could be easily modified by external circuitry to perform mathematical operations, such as constant gain multipliers, summing, subtraction, integration, and differentiation. With the advent of solid-state technology, op-amps have become highly reliable, miniaturized, temperature-stabilized, and consistently predictable in performance[1]. They now figure as fundamental building blocks in basic amplification and signal conditioning, in active filters, function generators, and switching circuits.

### Inverting amplifier

The inverting amplifier of Figure 10-1 has its noninverting input connected to ground or common. A signal is applied through input resistor $R_1$, and negative current feedback is implemented through feedback resistor $R_F$. Output $u_o$ has polarity opposite that of input $u_i$.

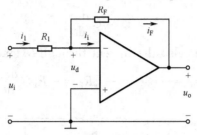

Figure 10-1　Inverting amplifier

### Summing amplifier

A popular use of an op-amp is as a summing amplifier. Figure 10-2 shows the connection. Figure 10-2 is formed by adding parallel inputs to the inverting input. Its output is a weighted sum of the inputs, but inverted in polarity. In an ideal op-amp, there is no limit to the number of inputs; however, the gain is reduced as inputs are added to a practical op-amp.

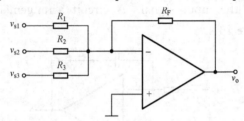

Figure 10-2　Summing amplifier

We use the principle of superposition. With $v_{s2}=v_{s3}=0$, the current in $R_1$ is not affected by the presence of $R_2$ and $R_3$, since the inverting node is a virtual ground. Hence, the output voltage due to $v_{s1}$ is $v_{o1}=-(R_F/R_1)v_{s1}$.

Similarly, $v_{o2}=-(R_F/R_2)v_{s2}$ and $v_{o3}=-(R_F/R_3)v_{s3}$. Then, by superposition,

$$v_o = v_{o1} + v_{o2} + v_{o3} = -R_F\left(\frac{v_{s1}}{R_1} + \frac{v_{s2}}{R_2} + \frac{v_{s3}}{R_3}\right)$$

$$\text{PSRR} = \frac{\Delta V_{DD}}{\Delta V_o}$$

$$A_V = \frac{V_o/V_i}{V_o/V_{DD}}$$

### Differentiation amplifier

The introduction of a capacitor into the input path of an op-amp leads to time differentiation of the input signal. The circuit of Figure 10-3 represents the simplest inverting differentiator involving an op-amp. As such, the circuit finds limited practical use, and since high-frequency noise can produce a derivative whose magnitude is comparable to that of the signal. In practice, low-pass filtering is utilized to reduce the effects of noise.

Since the op-amp is ideal, $v_d \approx 0$, and the inverting terminal is a virtual ground. Consequently, $v_s$ appears across capacitor $C$:

$$i_s = C \frac{dv_s}{dt}$$

But the capacitor current is also the current through $R$ (since $i_i = 0$). Hence,

$$v_o = -i_F R_F = -i_s R_F = -R_F C \frac{dv_s}{dt}$$

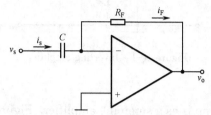

Figure 10-3  Differentiation amplifier

**Integrator amplifier**

The insertion of a capacitor in the feedback path of an op-amp results in an output signal that is a time integral of the input signal. A circuit arrangement for a simple inverting integrator is given in Figure 10-4.

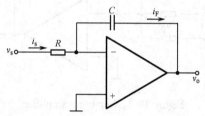

Figure 10-4  Integrator amplifier

If the op-amp is ideal, the inverting terminal is a virtual ground, and $v_s$ appears across $R$. Thus, $i_s = v_s/R$. But, with negligible current into the op-amp, the current through $R$ must also flow through $C$. Then

$$v_o = -\frac{1}{C}\int i_s \, dt = -\frac{1}{RC}\int v_s \, dt$$

# New Words

originally [əˈridʒənəli] *adv.* 最初，原先
substation [ˈsʌbsteiʃən] *n.* 分站，分所，变电站
predictable [priˈdiktəbəl] *adj.* 可预言的
miniaturize [ˈminiətʃəraiz] *vt.* 使小型化
noninverting [ˈnɔninˈvəːtiŋ] *n.* 非反相
feedback [ˈfiːdbæk] *n.* [无]回授，反馈，反应
superposition [ˌsjuːpəpəˈziʃən] *n.* 重叠，重合，叠合

## Phrases & Expressions

such as 例如……,像这种的
add to 增加
as such 同样地,同量地

## Technical Terms

feedback resistor 反馈电阻
high-pass filter 高阶滤波器
virtual ground 虚地
active filter 有源滤波器
switching circuit 开关电路

## Notes

1. With the advent of solid-state technology, op-amps have become highly reliable, miniaturized, temperature-stabilized, and consistently predicatable in performance.

随着固态技术的发展,运算放大器已经变得高度可靠、微型化、温度性能稳定,且性能是可预测的。

## Exercises

1. Keywords

In the article, there are some important words which are the soul of this paper. After reading this paper, we can find some words to stand for this article. Now please find out key words of this paper.

2. Summary

After reading this paper, please write a summary about IC with 200~300 words.

# 科技英语知识 3：英文书信

英文书信可分为私人信件和事务信件两大类。私人信件又称非正式信件（informal letter），事务信件一般是正式信件（formal letter）。值得注意的是，给陌生人或重要的人写信也要使用正式信件。

**信封**

英文信函的信封由三个主要部分构成，其格式与中文信函不同：发信人的姓名、地址或单位名称放在信封的左上角，收信人的姓名、地址或单位名称通常居中，邮票置于信封的右上角。

书写英文信函的信封应注意：

1. 地址顺序由小到大，先写姓名或单位名称，再写门牌号和具体地点，有邮政编码的不要忘记加上，寄往国外的信函要写上国名。如果地址较长，可以分为几行，一般按并列式排列。在英式用法中，允许采用斜列式。

2. 为表示礼貌，收信人的姓名前通常冠以 Mr.、Mrs.、Miss、Ms. 等称谓。若知道收信人的头衔，则应在其姓名前加上头衔；若不知道收信人的姓名，则可以写上 To Whom It May Concern（致有关人士），或者直接写上公司或机构的名称。

**信笺**

信笺通常由下列部分组成：发信人的地址、日期、称呼、正文、结尾、署名。正式信函还包括收信人的姓名、地址或单位名称，放在称呼之上。一般来讲，私人信笺要手写，事务信笺要打印出来，但并不严格要求。

信笺各部分的书写要点如下：

1. 信笺内的姓名和地址的写法与信封上的基本类似。一般采用分列式，信笺的右上角先写上发信人的地址，再写上发信日期；也可以采用并列式，发信人的地址和发信日期都写在信笺的右边。收信人的姓名、地址或单位名称写在左边，其位置要低于发信人的地址和发信日期的位置。在英式用法中，地址的每行末尾都加逗号，最后一行加句号，美式用法中不需要。

2. 中英文信函的发信日期有不同的写法。中文信函的日期写在署名之后，英文信函的日期写在信笺的开始。在美式用法中，发信日期的书写顺序是月、日、年，英式用法中的书写顺序是月、日、年或日、月、年，如"April 16, 2010"。

3. 信笺中对收信人的称呼与信封上的大致相似，称呼的末尾加逗号或冒号。

4. 结束正文后，在署名前要添加结尾客套语，客套语的末尾加逗号。

5. 署名时先手写签名，然后在手写签名的下面再打印姓名。如果发信人为女性，与收信人又不相识，则可以在打印的姓名左边或右边用括号注上 Miss、Mrs. 或 Ms.。署名之后还可以添上本人的职称或头衔。

# Unit 5  Power

## Lesson 11  Buck-Boost Power Stage and Steady-State Analysis

### 1. Buck-boost power stage introduction

The three basic switching power supply[1] topologies in common use are the buck, boost, and buck-boost. These topologies are nonisolated, i. e. , the input and output voltages share a common ground. There are, however, isolated derivations of these nonisolated topologies. The power supply topology refers to how the switches, output inductor, and output capacitor are connected. Each topology has unique properties. These properties include the steady-state voltage conversion ratios, the nature of the input and output currents, and the character of the output voltage ripple. Another important property is the frequency response of the duty-cycle-to-output-voltage transfer function.

The buck-boost is a popular nonisolated, inverting power stage topology, sometimes called a step-up/down power stage. Power supply designers choose the buck-boost power stage because the output voltage is inverted from the input voltage, and the output voltage can be either higher or lower than the input voltage. The topology gets its name from producing an output voltage that can be higher (like a boost power stage) or lower (like a buck power stage) in magnitude than the input voltage. However, the output voltage is opposite in polarity from the input voltage. The input current for a buck-boost power stage is discontinuous or pulsating due to the power switch (Q1) current that pulses from zero to $I_L$ every switching cycle. The output current for a buck-boost power stage is also discontinuous or pulsating. This is because the output diode only conducts during a portion of the switching cycle. The output capacitor supplies the entire load current for the rest of the switching cycle.

This report describes steady state operation of the buck-boost converter in continuous-mode and discontinuous-mode operation with ideal waveforms given. The duty-cycle-to-output-voltage transfer function is given after an introduction of the PWM switch model.

Figure 11-1 shows a simplified schematic of the buck-boost power stage with a drive circuit block included. The power switch, Q1, is an n-channel MOSFET. The output diode is CR1. The inductor, $L$, and capacitor, $C$, make up the effective output filter. The capacitor ESR, $R_C$, (equivalent series resistance) and the inductor DC resistance, $R_L$, are included in the analysis. The resistor, $R$, represents the load seen by the power stage output.

During normal operation of the buck-boost power stage, Q1 is repeatedly switched on and off with the on- and off-times governed by the control circuit. This switching action gives rise to a train of pulses at the junction of Q1, CR1, and L. Although the inductor, L, is connected to the output capacitor, C, only when CR1 conducts, an effective L/C output filter is formed. It filters the train of pulses to produce a DC output voltage.

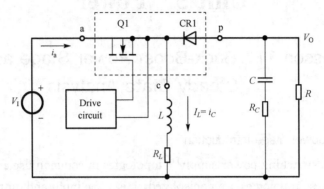

Figure 11-1  Buck-boost power stage schematic

## 2. Buck-boost stage steady-state analysis

A power stage can operate in continuous or discontinuous inductor current mode. Continuous inductor current mode is characterized by current flowing continuously in the inductor during the entire switching cycle in steady-state operation. Discontinuous inductor current mode is characterized by the inductor current being zero for a portion of the switching cycle. It starts at zero, reaches a peak value, and returns to zero during each switching cycle. The two different modes are discussed in greater detail later and design guidelines for the inductor value to maintain a chosen mode of operation as a function of rated load are given. It is very desirable for a converter to stay in one mode only over its expected operating conditions because the power stage frequency response changes significantly between the two different modes of operation[2].

For this analysis, an n-channel power MOSFET is used and a positive voltage, $V_{GS(ON)}$, is applied from the Gate to the Source terminals of Q1 by the drive circuit to turn ON the FET. The advantage of using an n-channel FET is its lower $R_{DS(ON)}$ but the drive circuit is more complicated because a floating drive is required. For the same die size, a p-channel FET has a higher $R_{DS(ON)}$ but usually does not require a floating drive circuit.

The transistor Q1 and diode CR1 are drawn inside a dashed-line box with terminals labeled a, p, and c. This is explained fully in the *Buck-Boost Power Stage Modeling* section.

## 3. Buck-boost steady-state continuous conduction mode analysis

The following is a description of steady-state operation in continuous conduction mode. The main goal of this section is to provide a derivation of the voltage conversion relationship for the continuous conduction mode buck-boost power stage. This is important because it shows how the output voltage depends on duty cycle and input voltage or conversely, how the duty cycle can be calculated based on input voltage and output voltage. Steady-state implies that the input voltage, output voltage, output load current, and duty-cycle are fixed and not varying. Capital letters are generally given to variable names to indicate a steady-state quantity.

In continuous conduction mode, the buck-boost converter assumes two states per switching cycle. The ON State is when Q1 is ON and CR1 is OFF. The OFF State is when Q1 is OFF and CR1 is ON. A simple linear circuit can represent each of the two states where the switches in the circuit are replaced by their equivalent circuit during each state. The circuit diagram for each of the two states is shown in Figure 11-2.

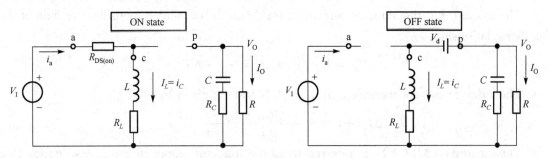

Figure 11-2 Buck-boost power stage states

The duration of the ON state is $DT_S = T_{ON}$ where $D$ is the duty cycle, set by the control circuit, expressed as a ratio of the switch ON time to the time of one complete switching cycle, $T_S$. The duration of the OFF state is called $T_{OFF}$. Since there are only two states per switching cycle for continuous conduction mode, $T_{OFF}$ is equal to $(1-D) T_S$. The quantity $(1-D)$ is sometimes called $D'$. These times are shown along with the waveforms in Figure 11-3.

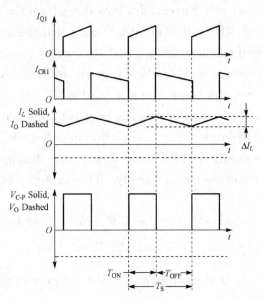

Figure 11-3 Continuous mode buck-boost power stage waveforms

Referring to Figure 11-2, during the ON state, Q1 presents a low resistance, $R_{DS(on)}$, from its drain to source and exhibits a small voltage drop of $V_{DS} = I_L R_{DS(on)}$. There is also a small voltage drop across the dc resistance of the inductor equal to $I_L R_L$. Thus, the input voltage, $V_I$, minus losses, $(V_{DS} = I_L R_L)$, is applied across the inductor, L. CR1 is OFF during this time because it is reverse biased. The inductor current, $I_L$, flows from the input source, $V_I$, through Q1 and to ground. During the ON state, the voltage applied across the inductor is constant and equal to $(V_I - V_{DS} - I_L R_L)$. Adopting the polarity convention for the current $I_L$ shown in Figure 11-2, the inductor current increases as a result of the applied voltage. Also, since the applied voltage is essentially constant, the inductor current increases linearly. This increase in inductor current during $T_{ON}$ is illustrated in Figure 11-3.

The amount that the inductor current increases can be calculated by using a version of the familiar relationship:

$$V_L = L \cdot \frac{di_L}{dt} \Rightarrow \Delta I_L = \frac{V_L}{L} \cdot \Delta T$$

The inductor current increase during the ON state is given by:

$$\Delta I_L(+) = \frac{V_I - (V_{DS} + I_L R_L)}{L} \cdot T_{ON}$$

This quantity, $\Delta I_L(+)$, is referred to as the inductor ripple current. Also notice that during this period, all of the output load current is supplied by the output capacitor, C.

Referring to Figure 11-2, when Q1 is OFF, it presents a high impedance from its drain to source. Therefore, since the current flowing in the inductor L cannot change instantaneously, the current shifts from Q1 to CR1. Due to the decreasing inductor current, the voltage across the inductor reverses polarity until rectifier CR1 becomes forward biased and turns ON. The voltage applied across L becomes $(V_O - V_d - I_L R_L)$ where the quantity, $V_d$, is the forward voltage drop of CR1. The inductor current, $I_L$, now flows from the output capacitor and load resistor combination through CR1 and to ground. Notice that the orientation of CR1 and the direction of current flow in the inductor means that the current flowing in the output capacitor and load resistor combination causes $V_O$ to be a negative voltage. During the OFF state, the voltage applied across the inductor is constant and equal to $(V_O - V_d - I_L R_L)$. Maintaining our same polarity convention, this applied voltage is negative (or opposite in polarity from the applied voltage during the ON time), because the output voltage $V_O$ is negative. Hence, the inductor current decreases during the OFF time. Also, since the applied voltage is essentially constant, the inductor current decreases linearly. This decrease in inductor current during $T_{OFF}$ is illustrated in Figure 11-3.

The inductor current decrease during the OFF state is given by:

$$\Delta I_L(-) = \frac{-(V_O - V_d - I_L R_L)}{L} \cdot T_{OFF}$$

This quantity, $\Delta I_L(-)$, is also referred to as the inductor ripple current.

n steady state conditions, the current increase, $\Delta I_L(+)$, during the ON time and the current decrease during the OFF time, $\Delta I_L(-)$, must be equal. Otherwise, the inductor current would have a net increase or decrease from cycle to cycle which would not be a steady state condition. Therefore, these two equations can be equated and solved for $V_O$ to obtain the continuous conduction mode buck-boost voltage conversion relationship:

Solving for $V_O$:

$$V_O = -\left[(V_I - V_{DS}) \cdot \frac{T_{ON}}{T_{OFF}} - V_d - I_L R_L \cdot \frac{T_{ON} + T_{OFF}}{T_{OFF}}\right]$$

And, substituting $T_S$ for $T_{ON} + T_{OFF}$, and using $D = T_{ON}/T_S$ and $(1-D) = T_{OFF}/T_S$, the steady-state equation for $V_O$ is:

$$V_O = -\left[(V_I - V_{DS}) \cdot \frac{D}{1-D} - V_d - \frac{I_L R_L}{1-D}\right]$$

Notice that in simplifying the above, $T_{ON} + T_{OFF}$ is assumed to be equal to $T_S$. This is true only for continuous conduction mode as we will see in the discontinuous conduction mode analysis.

An important observation should be made here: Setting the two values of $\Delta I_L$ equal to each other is precisely equivalent to *balancing the volt-seconds* on the inductor. The volt-seconds applied to the inductor is the product of the voltage applied and the time that the voltage is applied. This is the best way to calculate unknown values such as $V_O$ or $D$ in terms of known circuit parameters, and this method will be applied repeatedly in this paper. Volt-second balance on the inductor is a physical necessity and should be comprehended at least as well as Ohm's Law.

In the above equations for $\Delta I_L (+)$ and $\Delta I_L (-)$, the output voltage was implicitly assumed to be constant with no AC ripple voltage during the ON time and the OFF time. This is a common simplification and involves two separate effects. First, the output capacitor is assumed to be large enough that its voltage change is negligible. Second, the voltage due to the capacitor ESR is also assumed to be negligible. These assumptions are valid because the AC ripple voltage is designed to be much less than the DC part of the output voltage.

The above voltage conversion relationship for $V_O$ illustrates the fact that $V_O$ can be adjusted by adjusting the duty cycle, $D$. This relationship approaches zero as $D$ approaches zero and increases without bound as $D$ approaches 1. A common simplification is to assume $V_{DS}$, $V_d$, and $R_L$ are small enough to ignore. Setting $V_{DS}$, $V_d$, and $R_L$ to zero, the above equation simplifies considerably to:

$$V_O = -V_I \cdot \frac{D}{1-D}$$

A simplified, qualitative way to visualize the circuit operation is to consider the inductor as an energy storage element. When Q1 is on, energy is added to the inductor. When Q1 is off, the inductor delivers some of its energy to the output capacitor and load. The output voltage is controlled by setting the on-time of Q1. For example, by increasing the on-time of Q1, the amount of energy delivered to the inductor is increased. More energy is then delivered to the output during the off-time of Q1 resulting in an increase in the output voltage.

Unlike the buck power stage, the average of the inductor current is not equal to the output current. To relate the inductor current to the output current, referring to Figures 11-2 and 11-3, note that the inductor delivers current to the output only during the off state of the power stage. This current averaged over a complete switching cycle is equal to the output current because the average current in the output capacitor must be equal to zero.

The relationship between the average inductor current and the output current for the continuous mode buck-boost power stage is given by:

$$I_{L(\text{Avg})} \cdot \frac{T_{\text{OFF}}}{T_{\text{S}}} = I_{L(\text{Avg})}(1-D) = -I_{\text{O}}$$

or

$$I_{L(\text{Avg})} = \frac{-I_{\text{O}}}{(1-D)}$$

Another important observation is that the average inductor current is proportional to the output current, and since the inductor ripple current, $\Delta I_L$, is independent of output load current, the minimum and the maximum values of the inductor current track the average inductor current exactly[3]. For example, if the average inductor current decreases by 2A due to a load current decrease, then the minimum and maximum values of the inductor current decrease by 2A (assuming continuous conduction mode is maintained).

The forgoing analysis was for the buck-boost power stage operation in continuous inductor current mode. The next section is a description of steady-state operation in discontinuous conduction mode. The main result is a derivation of the voltage conversion relationship for the discontinuous conduction mode buck-boost power stage.

## New Words

invert [in'vɜːt] vt. 使……前后倒置,使反转
instantaneously [ˌinstən'teiniəsli] adv. 瞬时,立刻
negative ['negətiv] adj. 负数的,负的
constant ['kɒnstənt] n. 常数,常量;adj. 恒定的
opposite ['ɒpəzit] adj. 对立的
simplification [ˌsimplifi'keiʃn] n. 简化
deliver [di'livə(r)] vt. 传送,传递
visualize ['viʒuəlaiz] vt. 设想,假设
forgo [fɔː'gəu] v. 没有,放弃也行
derivation [ˌderi'veiʃn] n. 推导,导出

## Phrases & Expressions

substitute A for B    用 A 替代 B
in terms of    根据,就……而言
be much less than    远小于,远少于
be proportional to    与……成比例
be independent to    与……无关,脱离……而独立

## Technical Terms

drain [drein] n. 漏极

polarity [pə'lærəti] *n.* 极性
rectifier ['rektifaiə] *n.* 整流器
resistor [ri'zistə(r)] *n.* 电阻器
ohm [əum] *n.* 欧姆(电阻单位)
power stage 功率级,功率驱动器
control circuit 控制电路,控制回路
power conversion 能量转换,功率转换
ripple current 波纹电流
voltage drop 电压降
inductor current 电感电流
load current 负载电流
continuous condition mode 连续导通模式
output current 输出电流
energy storage element 能量储存单元
switching cycle 开关周期

## Notes

1. switching power supply：开关电源,是利用现代电力电子技术,控制开关管开通和关断的时间比例、维持稳定输出电压的一种电源。开关电源一般由脉冲宽度调制(PWM)控制 IC 和 MOSFET 构成。随着电力电子技术的发展和创新,开关电源技术也在不断地创新。目前,开关电源因其小型、轻量和高效率的特点被广泛应用于几乎所有的电子设备,是当今电子信息产业飞速发展不可缺少的一种电源。

2. It is very desirable for a converter to stay in one mode only over its expected operating conditions because the power stage frequency response changes significantly between the two different modes of operation.

在预期工作条件下仅保持一种模式,对转换器来说是很理想的,因为两种不同工作模式下功率级频率响应的变化是很显著的。

这个句子用到了 it is *adj.* for sb to do sth. 这种常用的句式,其含义是做 sth. 对 sb. 来说很 *adj.*。这里的 it 是形式主语,*adj.* 表示 do sth. 的属性,而 sb. 则跟 *adj.* 没有直接联系。该句还附加了一个由 because 引导的原因状语从句,用以说明保持同一模式的必要性。

3. Another important observation is that the average inductor current is proportional to the output current, and since the inductor ripple current, $\Delta I_L$, is independent of output load current, the minimum and the maximum values of the inductor current track the average inductor current exactly.

另一个重要的经验就是,电感电流的平均值与输出电流呈正比,因为电感波纹电流 $\Delta I_L$ 与输出负载电流无关,且电感电流的最大值和最小值精确地随电感电流的平均值而变化。

这个句子的框架结构是 ... observation is that ... and since the inductor ... 。主句是一

个从属连词宾语从句,由关系代词 that 引导,主语是 observation,谓语是 is;并由从属连词 since 引出一个原因状语从句,用来说明电感电流均值与输出电流成正比的原因。

# Lesson 12  Buck-Boost Steady-State Discontinuous Conduction Mode Analysis

We now investigate what happens when the load current is decreased and the conduction mode changes from continuous to discontinuous. Recall for continuous conduction mode, the average inductor current tracks the output current, i.e. if the output current decreases, then so does the average inductor current. In addition, the minimum and maximum peaks of the inductor current follow the average inductor current exactly.

If the output load current is reduced below the critical current level, the inductor current will be zero for a portion of the switching cycle. This should be evident from the waveforms shown in Figure 11-3, since the peak to peak amplitude of the ripple current does not change with output load current. In a buck-boost power stage, if the inductor current attempts to fall below zero, it just stops at zero (due to the unidirectional current flow in CR1) and remains there until the beginning of the next switching cycle. This operating mode is called discontinuous conduction mode. A power stage operating in discontinuous conduction mode has three unique states during each switching cycle as opposed to two states for continuous conduction mode. The inductor current condition where the power stage is at the boundary between continuous and discontinuous mode is shown in Figure 12-1. This is where the inductor current just falls to zero and the next switching cycle begins immediately after the current reaches zero. Note that the absolute values of $I_O$ and $I_{O(Crit)}$ are shown in Figure 12-1 because $I_O$ and $I_L$ have opposite polarities.

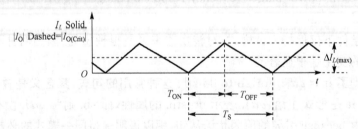

Figure 12-1  Boundary between continuous and discontinuous mode

Further reduction in output load current puts the power stage into discontinuous conduction mode. This condition is illustrated in Figure 12-2. The discontinuous mode power stage frequency response is quite different from the continuous mode frequency response and is given in the Buck-Boost Power Stage Modeling section. Also, the input to output relationship is quite different as shown in the following derivation.

To begin the derivation of the discontinuous conduction mode buck-boost power stage voltage conversion ratio, recall that there are three unique states that the converter assumes during discontinuous conduction mode operation[1]. The ON State is when Q1 is ON and CR1 is OFF. The OFF State is when Q1 is OFF and CR1 is ON. The IDLE state is when both Q1 and CR1 are OFF. The first two states are identical to those of the continuous mode case and the circuits of Figure 11-2 are applicable except that $T_{OFF} \neq (1-D)\, T_S$. The remainder of the switching cycle is the IDLE state. In addition, the DC resistance of the output inductor, the output diode forward voltage drop, and the power MOSFET ON-state voltage drop are all assumed to be small enough to omit.

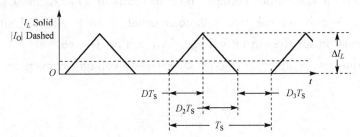

Figure 12-2  Discontinuous current mode

The duration of the ON state is $T_{ON} = DT_S$ where $D$ is the duty cycle[2], set by the control circuit, expressed as a ratio of the switch ON time to the time of one complete switching cycle, $T_S$. The duration of the OFF state is $T_{OFF} = D_2 T_S$. The IDLE time is the remainder of the switching cycle and is given as $T_S - T_{ON} - T_{OFF} = D_3 T_S$. These times are shown with the waveforms in Figure 12-3.

Without going through the detailed explanation as before, the equations for the inductor current increase and decrease are given below.

The inductor current increase during the ON state is given by:

$$\Delta I_L(+) = \frac{V_I}{L} \cdot T_{ON} = \frac{V_I}{L} \cdot DT_S = I_{PK}$$

The ripple current magnitude, $\Delta I_L(+)$, is also the peak inductor current, $I_{PK}$ because in discontinuous mode, the current starts at zero each cycle.

The inductor current decrease during the OFF state is given by:

$$\Delta I_L(-) = \frac{-V_O}{L} \cdot T_{OFF} = \frac{-V_O}{L} \cdot D_2 T_S$$

As in the continuous conduction mode case, the current increase, $\Delta I_L(+)$, during the ON time and the current decrease during the OFF time, $\Delta I_L(-)$, are equal. Therefore, these two equations can be equated and solved for $V_O$ to obtain the first of two equations to be used to solve for the voltage conversion ratio:

$$V_O = -V_I \cdot \frac{T_{ON}}{T_{OFF}} = -V_I \cdot \frac{D}{D_2}$$

Now we calculate the output current (the output voltage $V_O$ divided by the output load $R$). It is the average over one switching cycle of the inductor current during the time when CR1 conducts ($D_2 T_S$).

$$\frac{V_O}{R} = I_O = \frac{1}{T_S} \cdot \left(\frac{-I_{PK}}{2} D_2 \cdot T_S\right)$$

Now, substitute the relationship for $I_{PK}(\Delta I_L(+))$ into the above equation to obtain:

$$\frac{V_O}{R} = I_O = \frac{1}{T_S}\left[\frac{1}{2} \times (-1) \times \left(\frac{V_I}{L} DT_S\right) D_2 T_S\right]$$

$$\frac{V_O}{R} = \frac{-V_I D D_2 T_S}{2L}$$

We now have two equations, the one for the output current ($V_O$ divided by $R$) just derived and the one for the output voltage, both in terms of $V_I$, $D$, and $D_2$. We now solve each equation for $D_2$ and set the two equations equal to each other. Using the resulting equation, an expression for the output voltage, $V_O$, can be derived.

The discontinuous conduction mode buck-boost voltage conversion relationship is given by:

$$V_O = -V_I \cdot \frac{D}{\sqrt{K}}$$

Where $K$ is defined as

$$K = \frac{2L}{RT_S}$$

The above relationship shows one of the major differences between the two conduction modes. For discontinuous conduction mode, the voltage conversion relationship is a function of the input voltage, duty cycle, power stage inductance, the switching frequency, and the output load resistance. For continuous conduction mode, the voltage conversion relationship is only dependent on the input voltage and duty cycle.

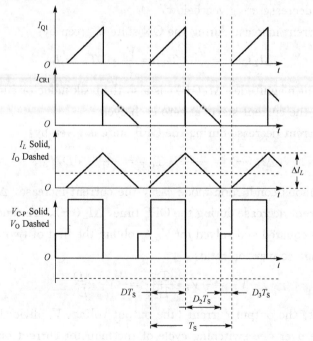

Figure 12-3 Discontinuous mode buck-boost converter waveforms

In typical applications, the buck-boost power stage is operated in either continuous conduction mode or discontinuous conduction mode. For a particular application, one conduction mode is chosen and the power stage is designed to maintain the same mode. The next section gives inductance relationships for the power stage that allow it to operate in only one conduction mode, given ranges for input and output voltage and output load current[3].

## New Words

recall [ri'kɔ:l] vt. 回想，召回
investigate [in'vestigeit] vt. 调查，研究
except [ik'sept] vt. 把……除外，除……外
remainder [ri'meində(r)] n. 剩余部分，剩余时间
reduction [ri'dʌkʃn] n. 减少，降低
derive [di'raiv] v. 得到，源于

## Phrases & Expressions

as opposed to    与……相对
be identical to  与……相同，与……相符
in addition      另外，除此之外

## Technical Terms

peak [pi:k] n. 峰值
unidirectional current    单向电流
critical current   临界电流
discontinuous condition mode   非连续导通模式
voltage conversion ratio   电压转换比
duty cycle    占空比

## Notes

1. To begin the derivation of the discontinuous conduction mode buck-boost power stage voltage conversion ratio, recall that there are three unique states that the converter assumes during discontinuous conduction mode operation.

为了开始推导非连续导通模式 buck-boost 功率级的电压转换比,先回想一下在非连续导通模式下工作的变压器的三种独特状态。

这个句子的正常语序应该是 ... recall that there are ... to begin the derivation of ...。目的状语从句 to begin the derivation of ... 在本句中被前置,意在突出 recall 的目的是为推导非连续导通模式的电压转换比做准备。

2. duty cycle：占空比,它在电信领域的含义是,在一串理想的脉冲周期序列(如方波)中,

正脉冲的持续时间与脉冲总周期的比值。简单来讲，占空比就是高电平在一个周期之内所占的时间比例。

3. The next section gives inductance relationships for the power stage that allow it to operate in only one conduction mode, given ranges for input and output voltage and output load current.

下一节给出了功率级的电感关系，它在已知给定的输入电压、输出电压和输出负载电流的范围内，仅允许功率级工作在一种导通模式下。

这个句子的框架结构是 ... section gives inductance relationships ... that allow it to ... given ranges for ... 。代词 that 引导一个定语从句，其先行词是 inductance relationships。在这个从句中又包含一个由 given 引出的条件状语从句，以限定功率级工作模式的选定条件。

# Lesson 13  Flyback Power Stage

A transformer-coupled variation of the traditional buck-boost power stage is the flyback power stage. This power stage operates like the traditional buck-boost power stage except that the single winding inductor is replaced with a two (or more) winding coupled inductor. The power switch, Q1 in Figure 13-1, applies the input voltage to the primary side ($L_{PRI}$) of the coupled inductor. Energy is stored until Q1 is turned off. Energy is then delivered to the output capacitor and load resistor combination from the secondary side ($L_{SEC}$) of the coupled inductor through the output diode CR1. This power stage provides electrical isolation of the input voltage from the output voltage. Besides providing electrical isolation, the isolation transformer can step-down (or step-up) the input voltage to the secondary. The transformer turns ratio can be designed so that reasonable duty cycles are obtained for almost any input voltage/output voltage combination, thus avoiding extremely small or extremely high duty cycle values.

The flyback power stage also eliminates two characteristics which sometimes make the standard buck-boost power stage unattractive; i. e. , the output voltage is opposite in polarity from the input voltage and the power switch requires a floating drive. Besides providing isolation, the coupled inductor secondary can be connected to produce an output voltage of either positive or negative polarity. In addition, since the power switch is in series with the primary of the coupled inductor, the power switch can be connected so that the source is ground referenced instead of connecting the drain to the input voltage as in the standard buck-boost power stage[1].

The flyback power stage is very popular in 48-V input telecom applications and 110-V AC or 220V AC off-line applications for output power levels up to approximately 50 watts. The exact power rating[2] of the flyback power stage, of course, is dependent on the input voltage/output voltage combination, its operating environment and many other factors. Additional output voltages can be generated easily by simply adding another winding to the coupled inductor along with an output diode and output capacitor. Obtaining multiple output voltages from a single power stage is another advantage of the flyback power stage.

A simplified schematic of the flyback power stage with a drive circuit block included is shown in Figure 13-1. In the schematic shown, the secondary winding of the coupled inductor is connected to produce a positive output voltage. The power switch, Q1, is an n-channel MOSFET. The diode, CR1, is usually called the output diode. The secondary inductance, $L_{SEC}$, and capacitor, $C$, make up the output filter. The capacitor $C_{SR}$, $R_C$ (equivalent series resistance[3]) is not included. The resistor, $R$, represents the load seen by the power supply output.

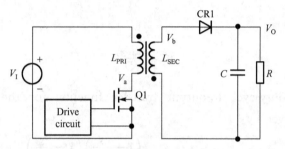

Figure 13-1  Flyback power stage schematic

The important waveforms for the flyback power stage operating in DCM are shown in Figure 13-2.

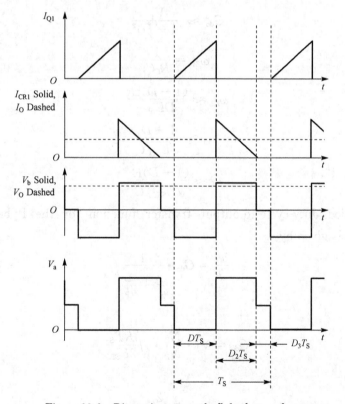

Figure 13-2  Discontinuous mode flyback waveforms

The simplified voltage conversion relationship for the flyback power stage operating in CCM (ignoring parasitics) is given by:

$$V_O = V_I \cdot \frac{N_S}{N_P} \cdot \frac{D}{1-D}$$

The simplified voltage conversion relationship for the flyback power stage operating in DCM (ignoring parasitics) is given by:

$$V_O = V_I \cdot \frac{N_S}{N_P} \cdot \frac{D}{\sqrt{K}}$$

Where $K$ is defined as:

$$K = \frac{2L_{SEC}}{RT_S}$$

The simplified duty-cycle-to-output transfer function for the flyback power stage operating in CCM is given by:

$$\frac{\hat{v}_O}{\hat{d}}(s) = G_{do} \cdot \frac{N_S}{N_P} \cdot \frac{\left(1+\frac{s}{\omega_{Z1}}\right)\left(1-\frac{s}{\omega_{Z2}}\right)}{1+\frac{s}{\omega_O Q}+\frac{s^2}{\omega_O^2}}$$

Where

$$G_{do} \approx \frac{V_I}{(1-D)^2}$$

$$\omega_{Z1} = \frac{1}{R_C C}$$

$$\omega_{Z2} \approx \frac{(1-D)^2 R}{DL_{SEC}}$$

$$\omega_O \approx \frac{1-D}{\sqrt{L_{SEC} C}}$$

$$Q \approx \frac{(1-D)R}{\sqrt{L_{SEC}/C}}$$

The simplified duty-cycle-to-output transfer function for the flyback power stage operating in DCM is given by:

$$\frac{\hat{v}_O}{\hat{d}} = G_{do} \cdot \frac{1}{1+\frac{s}{\omega_p}}$$

Where:

$$G_{do} = V_I \cdot \frac{N_S}{N_P} \cdot \sqrt{\frac{RT_S}{2L_{SEC}}}$$

and

$$\omega_p = \frac{2}{RC}$$

## New Words

combination [ˌkɒmbiˈneiʃn] *n.* 组合,联合体
besides [biˈsaidz] *prep.* 除……之外
secondary [ˈsekəndri] *adj.* 次等的,副的
reasonable [ˈriːznəbl] *adj.* 合理的,适当的
unattractive [ˌʌnəˈtræktiv] *adj.* 令人讨厌的,没意思的
eliminate [iˈlimineit] *vt.* 消除,排除
telecom [ˈteləkɔm] *n.* 电信

## Phrases & Expressions

in series with    与……串联
instead of    (用……)代替,而不是

## Technical Terms

watt [wɔt] *n.* 瓦特(功率单位)
inductance [inˈdʌktəns] *n.* 电感,电感值,感应
parasitics [ˌpærəˈsitiks] *n.* 寄生效应,寄生现象
flyback power stage    反向功率级
winding inductor    绕组电感线圈
primary side    主线圈
secondary side    副线圈
electrical isolation    电气隔离
operating environment    运行环境
power rating    额定功率
equivalent series resistance    等效串联电阻

## Notes

1. In addition, since the power switch is in series with the primary of the coupled inductor, the power switch can be connected so that the source is ground referenced instead of connecting the drain to the input voltage as in the standard buck-boost power stage.

此外,由于功率开关与耦合电感的主线圈串联,因此应连接功率开关,以便使源极成为参考地,而不是像标准的 buck-boost 功率级那样,把漏极连接到输入电压。

这个句子的结构是 since the power … the power switch can be … so that the source …。其中包含了一个由从属连词 since 引导的原因状语从句,这类从句通常放在主句之前;主句之后紧

接的是一个由 so that 引导的目的状语从句,这类从句不能用逗号隔开,且不可在从句中使用情态动词。

2. power rating：额定功率,指用电器正常工作时的功率。它的值为用电器的额定电压乘以额定电流。用电器若实际功率大于额定功率,则可能会损坏;若实际功率小于额定功率,则可能无法运行。

3. equivalent resistance：等效电阻,几个连接起来的电阻所起的作用,可以用一个电阻代替,这个电阻就是那些电阻的等效电阻。也就是说,任何电路中的电阻,不论有多少只,都可等效为一个电阻来代替,而不影响原电路两端的电压和电路中电流强度的变化。这个等效电阻的值是由多个电阻经过等效串并联公式计算得出的。也可以说,将这一等效电阻代替原有的几个电阻后,对于整个电路的电压和电流不会产生任何影响,所以这个电阻就叫作电路中的等效电阻。

# Exercises

1. Keywords

In the article, there are some important words which are the soul of this paper. After reading this paper, we can find some words to stand for this article. Now please find out key words of this paper.

2. Summary

After reading this paper, please write a summary about filter.

# 科技英语知识 4：语法成分的转换

在翻译中，虽然总体上可以保持译文与原文中的主语、谓语、宾语等句子成分的对应，但由于英汉两种语系在思维方式和表达习惯上的差别，很多情况下要对原句做语法成分的必要转换，以克服翻译中遇到的表达上的障碍。

不论是意译、转译或词性转换的方法，还是下面讨论的语法成分转换的方法，都是建立在对原文透彻理解的基础上对其含义的灵活表达，而意译、词性转换在多数情况下必然引起语法成分的变化。

1. 在被动句中，主语作为动作的接受者而常常被转换成宾语来翻译，而宾语则按主语来翻译。

例：Considerable use has been made of these data.

这些资料得到了充分的利用。

2. 当 care、need、attention、improvement、emphasis 等名词或名词化结构做主语时，可采取将主语译为谓语的方式。

例：Care should be taken to protect the instrument from dust and dump.

应当注意保护仪器免受灰尘和潮湿。

3. 某些英语谓语动词，如 act、characterize、feature、behave、relate、conduct 等在翻译成汉语时往往要将词性转换为名词而在句中做主语。

例：Fuzzy control acts differently from conventional PID control.

模糊控制的作用不同于传统的比例积分微分控制。

4. 为了突出原句中的定语，可在翻译时将其转换为谓语或表语。

例：There is a large amount of energy wasted due to the fraction of commutator.

换向器引起的摩擦损耗了大量的能量。

5. 对于英语中有 to have a voltage of ...、to have a height of ... 等结构时，如按照字面意思直译为汉语，就会十分生硬或拗口，因此在翻译时通常把"电压"、"高度"等译作主语，同时对句中其他语法成分做相应的转换。

例：The output of transformer has a voltage of 10 kilovoltage.

变压器的输出电压为 10kV。

原文的写作是为了表达某一思路或阐述某一事物的性质、过程，而翻译的目的同样是表达这一思路或阐述这一事物的性质、过程，因此不必过分拘泥于原文的表达方式，片面追求忠实于原文而忽视英汉两种语言的内在差异。换句话说，根据实际情况需要，采取意译、转译、词性转换、语法成分转换等处理方法是极为必要的。

# Unit 6　Electronic Instruments

## Lesson 14　Oscilloscope

The word "oscilloscope" has evolved to describe any of a variety of electronic instruments used to observe, measure, or record transient physical phenomena and present the results in graphic form. Perhaps the popularity and usefulness of the oscilloscope spring from its exploitation of the relationship between vision and understanding. In any event, several generations of technical workers have found it to be an important tool in a wide variety of settings.

The prototypical oscilloscope produces a two-dimensional graph with the voltage presented at an input terminal plotted on the vertical axis and time plotted on the horizontal axis (Figure 14-1). Usually the graph appears as an illuminated trace on the screen of a cathode-ray tube (CRT) and is used to construct a useful model or representation of how the instantaneous magnitude of some quantity varies during a particular time interval [1]. The "quantity" measured is often a changing voltage in an electronic circuit. However, it could be something else, such as electric current, acceleration, light intensity, or any of many other possibilities, which has been changed into a voltage by a suitable transducer. The "time interval" over which the phenomenon is graphed may vary over many orders of magnitude, allowing measurements of events which proceed too quickly to be observed directly with the human senses. Instruments of current manufacture measure events occurring over intervals as short as tens of picoseconds ($10^{-12}$ s) or up to tens of seconds.

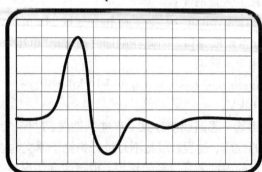

Figure 14-1　Voltage is plotted on the vertical axis and time horizontally
on the classic oscilloscope display

The measured quantities can be uniformly repeating or essentially nonrecurring. The most useful oscilloscopes have multiple input channels so that simultaneous observation of multiple phenomena is possible, allowing measurement of the time relationships of related events. For example, the time delay from clock to output of a D type logic flipflop can be measured (Figure 14-2). This type of flipflop copies the logic state on the input D to the output Q when the clock has a state change from low to high. The lower trace shows the flipflop clock, while the upper trace shows the resulting state change propagating to the Q output. The difference in the horizontal position of the two positive transitions indicates the time elapsed between the two events.

With the aid of an oscilloscope, rapidly time-varying quantities are captured as a static image and can be studied at a pace suitable for a human observer.

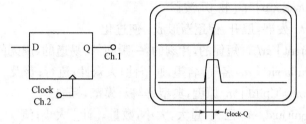

Figure 14-2 A time-interval measurement using a two-channel oscilloscope

The image can be preserved as a photograph or on paper with a plotter or printer. The record can be transferred into the memory of a digital computer to be preserved or analyzed.

The oscilloscope has for many years been used for a wide variety of measurements by engineers, scientists, and technicians. Many would describe it as among the most versatile and useful of the general-purpose electronic measuring tools, limited only by the imagination of those who use it.

An oscilloscope is ordinarily used to characterize the way a voltage varies with the passage of time. Therefore, any time-varying quantity which can be converted into a voltage also can be measured. Devices which change energy from one form to another are called "transducers" and are used for many purposes. For example, a loudspeaker changes electrical energy into sound waves and a microphone does the opposite conversion. The familiar glass tube thermometer uses heat energy to expand a column of mercury, giving a visual indication of the temperature of its surroundings.

The transducers useful in oscilloscope applications are those which convert energy into an electric voltage, and many such devices have been designed for adapted to this purpose. The microphone, although originally designed to allow the transmission of sound by telephone and radio, is now used in conjunction with an oscilloscope to explore in detail the duration and intensities of sound waves. An electric current probe, although just a special form of a transformer, is an example of a transducer designed specifically for use with an oscilloscope. The current to be sensed is linked through a magnetic core, and a voltage proportional to the input current is developed across the multiturn secondary winding and transmitted to the oscilloscope input for display.

Oscilloscopes are probably most often used for direct measurement of the transient voltage signals occurring within equipment that directly depends on electricity for its operation, such as computers, automation controls, telephones, radios, television, power supplies, and many others [2]. In equipment that functions by the use of rapidly changing electrical signals, pulses, or wave trains, the oscilloscope is useful for measuring the parameters that may be important in determining its correct operation: signal timing relationships, duration, sequence, rise and fall times, propagation delays, amplitudes, etc.[3]

## New Words

oscilloscope [ɔ'siləskəup] n. [物]示波器
evolve [i'vɔlv] vt. 发展,展开,使逐渐形成,使进化
transient ['trænziənt] adj. 短促的,片刻的,一瞬间的;易逝的,虚无的
popularity [ˌpɔpju'læriti] n. 名气,名望;通俗性;大众性;流行;普及
prototypical [ˌprəutə'tipikl] n. 原型;典型;样板,模范,标准
magnitude ['mæɡnitjuːd] n. 巨大,重大,大小,数量;[计] 大小;值
illuminate [i'ljuːmineit] vt. 照亮,照明,使光辉灿烂
transducer [trænz'djuːsə] n. 转换器;换流器,变频器;换能器,转换装置
imagination [iˌmædʒi'neiʃən] n. 想象,想象力,创造力
thermometer [θə'mɔmitə(r)] n. 体温计,温度计
mercury ['məːkjuri] n. 水银柱,水银剂
multiturn ['mʌltitəːn] adj. 多圈的,多匝的
function ['fʌŋkʃən] vi. 活动,运行,行使职责;n. 官能,职务,功能,函数

## Phrases & Expressions

spring from    出于,由……而来,由……造成
in any event   不论怎样,无论如何
too ... to ...    太……以至于不能……
as ... as ...    同……一样,如同
up to    多达,直到,至多
tens of    几十
so that    所以,因此,为使,以便
transfer into    转到
vary with    随……而变化
visual indication    可见指示(信号),目测,直观指示
in conjunction with    与……协力
transmit to    把……传给

## Technical Terms

vertical axis    纵轴

two-dimensional 二维的
cathode-ray （阴极发射出的）高速电子
measured quantity 测定量
simultaneous observation 同步观察
time relationship 时间关系
flipflop 触发器；触发电路，双稳态多谐振荡器
logic state 逻辑状态
time-varying 随时间变化，时变
current probe 电流探针
magnetic core 磁芯
secondary winding 复卷绕组
wave train [电]波列

## Notes

1. Usually the graph appears as an illuminated trace on the screen of a cathode-ray tube (CRT) and is used to construct a useful model or representation of how the instantaneous magnitude of some quantity varies during a particular time interval.

通常在阴极射线管（CRT）屏幕上，图形以明亮的痕迹显示，这个图形可以用来构建一个有用的模型，或者表示在特定扫描时间内一些量的瞬态值是如何变化的。

2. Oscilloscopes are probably most often used for direct measurement of the transient voltage signals occurring within equipment that directly depends on electricity for its operation, such as computers, automation controls, telephones, radios, television, power supplies, and many others.

示波器大概是最常用的直接测量瞬态电压信号的设备，这些信号产生于直接依靠电力运转的设备，如计算机、自动控制仪器、电话机、收音机、电视机、电源和其他设备。

3. In equipment that functions by the use of rapidly changing electrical signals, pulses, or wave trains, the oscilloscope is useful for measuring the parameters that may be important in determining its correct operation: signal timing relationships, duration, sequence, rise and fall times, propagation delays, amplitudes, etc.

在通过使用快速改变电信号、脉冲或者波列运行的设备中，示波器用来测量决定这些设备正确操作的重要参数：信号的时序关系、持续时间、顺序、上升和下降时间、传输延迟、振幅等。

## Lesson 15  Function Generators

A function generator is a piece of electronic test equipment or software used to generate electrical waveforms. These waveforms can be either repetitive, or single-shot in which case some kinds of triggering source are required (internal or external).

Another type of function generator is a sub-system that provides an output proportional to some mathematical function of its input; for example, the output may be proportional to the square root of the input. Such devices are used in feedback control systems and in analog computers. Analog function generators usually generate a triangle waveform as the basis for all of its other outputs. The triangle is generated by repeatedly charging and discharging a capacitor from a constant current source. This produces a linearly ascending or descending voltage ramp. As the output voltage reaches upper and lower limits, the charging and discharging are reversed using a comparator, producing the linear triangle wave. By varying the current and the size of the capacitor, different frequencies may be obtained.

A 50% duty cycle square wave is easily obtained by noting whether the capacitor is being charged or discharged, which is reflected in the current switching comparator's output. Most function generators also contain a non-linear diode shaping circuit that can convert the triangle wave into a reasonably accurate sine wave. It does so by rounding off the hard corners of the triangle wave in a process similar to clipping in audio systems.

The type of output connector from the device depends on the frequency range of the generator. A typical function generator can provide frequencies up to 20 MHz and uses a BNC connector, usually requiring a 50Ω or 75Ω termination. Specialized RF generators are capable of gigahertz frequencies and typically use N-type output connectors.

Function generators, like most signal generators, may also contain an attenuator, various means of modulating the output waveform, and often the ability to automatically and repetitively "sweep" the frequency of the output waveform (by means of a voltage-controlled oscillator) between two operator-determined limits [1]. This capability makes it very easy to evaluate the frequency response of a given electronic circuit. Some function generators can also generate white or pink noise.

More advanced function generators use Direct Digital Synthesis (DDS) to generate waveforms. Arbitrary waveform generators use DDS to generate any waveform that can be described by a table of amplitude values.

### Triangle wave

A triangle wave is a non-sinusoidal waveform named for its triangular shape. Figure 15-1 shows a band limited triangle ware pictured in the time domain and frequency domain.

Like a square wave, the triangle wave contains only odd harmonics. However, the higher harmonics roll off much faster than in a square wave.

It is possible to approximate a triangle wave with additive synthesis by adding odd harmonics of the fundamental, multiplying every $(4n-1)$ th harmonic by $-1$ (or changing its phase by $\pi$), and rolling off the harmonics by the inverse square of their relative frequency to the fundamental [2]. This infinite Fourier series converges to the triangle wave:

$$x_{\text{triangle}}(t) = \frac{8}{\pi^2} \sum_{k=0}^{\infty} (-1)^k \frac{\sin[2\pi(2k+1)ft]}{(2k+1)^2}$$

$$= \frac{8}{\pi^2} [\sin(2\pi ft) - \frac{1}{9}\sin(6\pi ft) + \frac{1}{25}\sin(10\pi ft) + \ldots ]$$

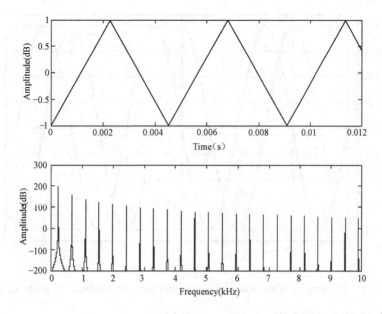

Figure 15-1  A bandlimited triangle wave pictured in the time domain (top) and frequency domain (bottom), the fundamental is at 220 Hz

## Sine wave

The sine wave or sinusoid is a function that occurs often in mathematics, music, physics, signal processing, audition, electrical engineering, and many other fields. Its most basic form is

$$y(t) = A\sin(\omega t + \theta)$$

which describes a wavelike function of time ($t$) with:
- peak deviation from center $= A$ (aka amplitude);
- angular frequency $\omega$, (radians per second);
- phase $= \theta$.

When the phase is non-zero, the entire waveform appears to be shifted in time by the amount $\theta/\omega$ seconds. A negative value represents a delay, and a positive value represents a "head-start".

The sine wave is important in physics because it retains its waveshape when added to another sine wave of the same frequency and arbitrary phase. It is the only periodic waveform that has this property. This property leads to its importance in Fourier analysis and makes it acoustically unique. Figure 15-2 shows the graphs of the sine and cosine functions.

## Sawtooth wave

The sawtooth wave (or saw wave) is a kind of non-sinusoidal waveform. It is named a sawtooth based on its resemblance to the teeth on the blade of a saw.

The convention is that a sawtooth wave ramps upward and then sharply drops. However, there are also sawtooth waves in which the wave ramps downward and then sharply rises. The latter type of sawtooth wave is called a "reverse sawtooth wave" or "inverse sawtooth wave". As audio signals, the two orientations of sawtooth wave sound identical.

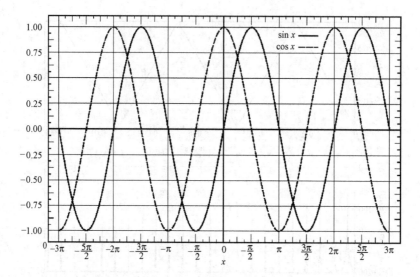

Figure 15-2　The graphs of the sine and cosine functions are sinusoids of different phases

The piecewise linear function

$$x(t) = t - \text{floor}(t)$$

based on the floor function of time $t$, is an example of a sawtooth wave with period 1.

A more general form, in the range $-1$ to $1$, and with period $a$, is

$$x(t) = 2\left[\frac{t}{a} - \text{floor}\left(\frac{t}{a} + \frac{1}{2}\right)\right]$$

This sawtooth function has the same phase as the sine function.

A sawtooth wave's sound is harsh and clear and its spectrum contains both even and odd harmonics of the fundamental frequency. Because it contains all the integer harmonics, it is one of the best waveforms to use for synthesizing musical sounds, particularly bowed string instruments like violins and cellos, using subtractive synthesis.

A sawtooth can be constructed using additive synthesis. The infinite Fourier series

$$x_{\text{sawtooth}}(t) = \frac{2}{\pi}\sum_{k=1}^{\infty}\frac{\sin(2\pi kft)}{k}$$

converges to an inverse sawtooth wave. A conventional sawtooth can be constructed using

$$x_{\text{sawtooth}}(t) = -\frac{2}{\pi}\sum_{k=1}^{\infty}\frac{\sin(2\pi kft)}{k}$$

In digital synthesis, these series are only summed over k such that the highest harmonic, $N_{\text{max}}$, is less than the Nyquist frequency[3] (half the sampling frequency). This summation can generally be more efficiently calculated with a Fast Fourier transform[4]. If the waveform is digitally created directly in the time domain using a non-bandlimited form, such as $y = x - \text{floor}(x)$, infinite harmonics are sampled and the resulting tone contains aliasing distortion.

A bandlimited sawtooth wave pictured in the time domain (top) and frequency domain (bottom) is shown in Figure 15-3.

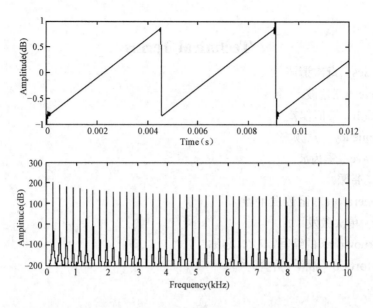

Firgure 15-3  A bandlimited sawtooth wave pictured in the time domain (top) and frequency domain (bottom), the fundamental is at 220Hz

## New Words

proportional [prə'pɔːʃnl] *adj.* （成）比例的；相称的,平衡的,调和的（to）
clip [klip] *vt.* 剪去,剪短,修剪；剪取,删削；削减
gigahertz ['giəhəːts] *n.* [频率单位]千兆赫（兹）
ramp [ræmp] *n.* 倾斜装置,斜轨；坡道,斜路；(筑城)斜坡
reasonably ['riːzənəbli] *adj.* 合理的；明白道理的,懂道理；有理性的
attenuator [ə'tenjueitə(r)] *n.* 衰减器
sinusoid ['sainəsɔid] *n.* 正弦曲线,窦状隙
phase [feiz] *n.* 时期,局面,方面,位相,相,阶段
acoustical [ə'kuːstikəl] *adj.* 听觉的,声学的
resemblance [ri'zembləns] *n.* 相似处,类似,肖像
orientation [ˌɔːrien'teiʃən] *n.* 定方位,适应,向东方；[计]方向
harsh [hɑːʃ] *adj.* 粗糙的,刺耳的,严厉的
spectrum ['spektrəm] *n.* 光谱,范围；[无]（射频，无线电信号）频谱
cello ['tʃeləu] *n.* 大提琴
tone [təun] *n.* 音调,音质,语调,语气

## Phrases & Expressions

by means of    依靠
a kind of    一种,一类

## Technical Terms

odd harmonics 奇次谐波
triangle wave 三角波
peak deviation 峰值偏差
angular frequency 角频率
sawtooth wave 锯齿波
bandlimite 带限
function generators 函数发生器
single-shot 单次激发
aka also known as 的缩写，又称，亦称
aliasing distortion 混淆（折叠）失真

## Notes

1. Function generators, like most signal generators, may also contain an attenuator, various means of modulating the output waveform, and often the ability to automatically and repetitively "sweep" the frequency of the output waveform (by means of a voltage-controlled oscillator) between two operator-determined limits.

像大部分信号发生器一样，函数发生器还可以包含一个以各种方式调节输出波形的衰减器，且可以经常在操作人员确定的两个限制值之间自动地重复"扫描"输出波形的频率（通过电压控制的振荡器）。

2. It is possible to approximate a triangle wave with additive synthesis by adding odd harmonics of the fundamental, multiplying every $(4n-1)$th harmonic by $-1$ (or changing its phase by $\pi$), and rolling off the harmonics by the inverse square of their relative frequency to the fundamental.

通过用基波的奇次谐波进行叠加逼近产生三角波，每 $4n-1$ 次谐波乘以 $-1$（或者通过乘以 $\pi$ 改变它的相位），然后通过基波相对频率平方的倒数求出谐波。

3. Nyquist frequency：奈奎斯特频率，是离散信号系统采样频率的一半，因哈里·奈奎斯特（Harry Nyquist）或奈奎斯特-香农采样定理得名。香农采样定理指出，只要离散系统的奈奎斯特频率高于采样信号的最高频率或带宽，就可以避免混叠现象。

4. Fast Fourier transform：快速傅里叶变换（FFT），是离散傅里叶变换的快速算法，也可用于计算离散傅里叶变换的逆变换。快速傅里叶变换有广泛的应用，如数字信号处理、计算大整数乘法、求解偏微分方程等。

# Lesson 16   Computer-Based Test Instruments

Personal-computer-based test instruments have fast become the quick and easy way for companies of all sizes to realize the advantages of computer — aided testing(CAT) without incurring the enormous expense of program development and main frame time[1]. Such instruments are able to meet the goals of reduced labor cost, increased productivity, and the elimination of human error in reading and processing measurements, by utilizing the power of software to perform many of the functions traditionally done by workbenches loaded with hardware[2].

The general design of these instruments is around a desktop personal computer, such as the IBM PC or the IBM PC AT. PC instruments, as they have come to be known, are categorized as being either internal or external.

### 1. Internal Adapters

PC instruments designed for internal use are fabricated on one or more computer adapter boards. These adapter boards are physically identical to the video and I/O adapter cards required by the computer for normal operation. To install the instrument into the computer, the card is simply plugged into an available expansion slot on the motherboard, and the system is powered up as usual. Use of the test instrument is then controlled by software from the computer's keyboard. There are no knobs, switches, or indicators available to the user. The entire operation of the instrument is through the computer and its interfaces.

An example of an internal-adapter PC-test device is an analytical oscilloscope from R. C. Electronics called the COMPUTERSCOPE-IND IS-16. The IS-16 Data Acquisition package consists of a 16-channel analog-to-digital conversion board, an external instrument interface box, and appropriate software.

In operation, the IS-16 offers a 1-MHz aggregate sampling rate capability on 16 individual input channels with 14-bit resolution at input voltages within the range of $-10V$ to $+10V$. Fully automated keystroke commands (which programmers call hotkeys) provide the user with control over all features of the instrument, including channel selection, trigger control (internal to any channel, external, $+/-$ level or slop), sampling rate, and memory buffer size from 1KB to 64KB. In effect, the hotkeys are designated keys on the computer keyboard that act like the knobs and levers found on an oscilloscope's front panel.

Beyond the simulated mechanical aspects of an oscilloscope. The IS-1 6 employs a ring buffer that allows the capture of data in pretrigger intervals of virtually any length. Software commands permit timerbase expansion and contraction, left and right scrolling, independent vertical gain adjustment, and waveform storage and retrieval.

It is this last feature, waveform storage and retrieval, that sets the PC-based IS-16 oscilloscope apart from its mechanical counterpart. Input measurements can be stored in computer files or temporary computer memory (RAM) for archival purposes or further processing. The fact that the dynamic input becomes static in nature provides the user with the ability to modify the contents of an existing file in many useful ways, So that subsequent analysis of the signal may be effectively performed. The entire operation is done in software, following the acquisition of the input signal (Figure 16-1).

Figure 16-1 Computer diskettes save measured raw data for subsequent processing

The program begins by creating a verbatim copy of the input signal in a computer data file, using strings of binary words to represent instantaneous events and values in real time. Once saved, the entire measurement procedure can be duplicated down to the last detail by simply playing back the file record into the computer, in much the same way music is forever captured on a record disk or magnetic tape and played back upon demand. Each channel of data may be saved and processed separately by this method.

The saved data may now be processed by the computer's software to derive measurements not initially performed by the PC instrument. Table 16-1 gives a brief summary of the procedures that may be performed on the raw data. When one realizes that all these measurements can be made after the fact, the implications are staggering. One-time events that are elusive to conventional measurement can be done routinely in the leisure of an office setting by simply running the recorded data through the computer one more time using the appropriate software selection.

Furthermore, the results can build one on another. For example, in the first pass, noise and amplification factors may be adjusted, resulting in a data file that can now be averaged, integrated, or differentiated to see the effects of each, without ever having to change the physical setup of the test instrument or device under test.

Table 16-1 Summary of Computerscope IS-16 computer-aided measurements

**GLOSSARY**

**Signal averaging**: Provides greater resolution by averaging the results over a user-specified period of time.

**Full-wave rectification**: Data points with voltage values below a user specified threshold are converted to voltage values above the threshold, relative to the initial difference in absolute magnitude.

**Data inversion**: The voltage value for each recorded data point is inverted in its relationship to a user-specified threshold.

**Gain adjustment**: Data points may be reassigned voltage values according to a new gain factor.

**Integration**: Channels may be integrated individually with a user-specified time constant, the results of which are adjusted in their temporal relationship so that no time lag is introduced.

**Differentiation**: The analog waveform is converted to its first derivative.

**DC offset adjustment**: The value for each recorded data point is increased or reduced by a user specified constant, thus adjusting baseline voltage.

**Algebraic functions**: Each waveform may be added, subtracted, multiplied, or divided by a user specified factor.

**Digital filtering**: Five differential digital frequency filtering functions ... low pass, high pass, band pass, band stop, and notch-permit the user to determine the cutoff frequencies and roll-off efficiency independently for each input channel.

**Three-point symmetric smoothing**: A three-point symmetric point provides the ability to attenuate high-frequency noise in the data.

## 2. External PC Instruments

Such performance is not limited to internal adapter devices or PC-based oscilloscopes. Similar results can be achieved using externally connected PC instruments.

In an externally connected PC instrument, the test instrument is housed in a conventional cabinet external to the computer's cabinet. Installed in the cabinet may be input-output connectors, selector switches, and maybe a LED or two. There is also a connector cable coming from the external instrument that plugs into the PC.

The connection between the external PC test instrument and the computer is in the form of an umbilical cord that feeds voltage, data, and control signals between the two devices. Without this connection, the test instrument is little more than a stand-alone device with features less than comparable to the instruments examined earlier in this book. When connected to the computer, however, it becomes a virtual powerhouse.

Inside the computer is an interface card that translates the signals coming from the external measuring instrument into digital pulses with voltage levels and timing requirements compatible with those on the computer bus (Figure16-2).

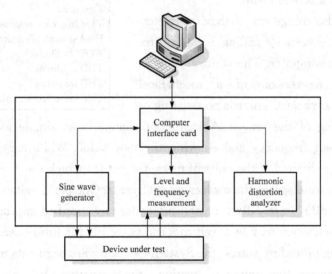

Figure 16-2 System One block diagram

An example of an external PC test instrument is the System One audio analyzer from Audio Precision. The System One is designed as a computer-interfaced audio test station capable of performing more than 36 standard performance tests on audio amplifiers, magnetic tape systems, and related audio components.

The System One unit contains of an ultra-low-distortion (0.0008% at midband) sine-wave generator, a level-detector and frequency-measurement amplifier, a harmonic distortion analyzer, and a computer interface card.

The System One has no controls or displays other than power on/off. Located on the front panel of the instrument are the audio connectors required to interface the System One with the device under test. All user — related functions of control, such as selection of test type, level adjustment, frequency output, and display are accomplished via the keyboard and display of an IBM PC-compatible personal computer.

Like the COMPUTERSCOPE-IND IS-16, all equipment and test functions are controlled by predefined keystrokes on the PC keyboard. Because most audio measurements are sets of measurements rather than an individual spot measurement, the keyboard is programmed to execute an entire series of measurements at the touch of a key, when requested. Generally, audio tests are performed using sweep frequencies that span beyond the range of the device under test. Some of the tests available are listed in Table 16-2.

The type of test performed, such as frequency response, is selected from a menu of standard tests. In these tests, the sweep range, rate, and amplitude are preset, according to SMPTE, DIN, CCIR, EIA, IHF, NAB, and other established audio standards. Each of these test patterns is recorded within the software of the System One application package for easy access through a screen menu.

The user may also change any of these parameters before initiating the sweep by calling up a control panel screen on the computer. The control panel is essentially a screen version of a mechanical quantities from the keyboard. There is no twisting of the knob or changing of the ranges. A frequency change is as simple as locating the line displaying the current frequency and entering the new value. While conventional hardware panel instruments are limited to the units of measurement for which the designer has the panel room and the meter scale space, the choices of the System One panel values can be changed to include volts, dBV, dBu, watts, dBm, or percentage for any specified impedance or function. If desired, the new parameters may be saved to a computer file for future use.

**Table 16-2  System One test functions**

| |
|---|
| Amplitude |
| Phase |
| Noise |
| Common mode rejection ratio |
| Crosstalk |
| Equalization curve |
| Frequency response |
| Frequency drift |
| Phase jitter |
| Tape noise |
| Tape frequency response |
| Head azimuth alignment |
| WOW & Flutter |
| THD amplitude |
| THD frequency |
| THD versus power output |

Once the testing procedure starts, the System One software begins its measurement of the unit under test. During the course of testing, the measurements are displayed on the computer's video monitor. The result is an X-versus-Y graph representation of the tests according to the response of the unit under test. Multiple line charts may be displayed by entering the proper values into the opening menus.

Test measurements are also saved to a computer file. As with the PC oscilloscope described above, these files may be played back for re-enactment of the tests performed or verification of test results. Unlike the COMPUTER-SCOPE, the System One does its testing in real time. Because of the nature of the tests involved, measurements are mor accurate when programmed to be performed on the unit under test at the time it is connected to test equipment than measurements derived from calculated values are. That is not to say that new measurements and conclusions cannot be derived from the data in the files created by the

System One, because the can. Therein lies the advantage of the PC-base test instrument over other types of testing devices.

## New Words

archival [ɑːˈkaivəl] *adj.* 关于档案的
desktop [ˈdesktɔp] *n.* 桌面
dynamic [daiˈnæmik] *adj.* 动态的
elusive [iˈluːsiv] *adj.* 难以捉摸的;不易记住的;逃避的
fabricate [ˈfæbrikeit] *vt.* 制造;装配;捏造
retrieval [riˈtriːvl] *n.* 收回,挽回;检索
slot [slɔt] *n.* 位置;狭槽,水沟

## Phrases & Expressions

adapter boards    适配器板
expansion slot    扩展槽

## Technical Terms

distortion [diˈstɔːʃn] *n.* 扭曲,变形;失真,畸变
keystroke [ˈkiːstrəuk] *n.* 键击,按键
motherboard [ˈmʌðəbɔːd] *n.* 底板,母板

## Notes

1. Personal-computer-based test instruments have fast become the quick and easy way for companies of all sizes to realize the advantages of computer — aided testing (CAT) without incurring the enormous expense of program development and main frame time.

基于PC的测试仪器已经迅速成为各类公司实现计算机辅助测试(CAT)的快捷方法,而不用在程序开发和结构搭建上花费大量的时间。

2. Such instruments are able to meet the goals of reduced labor cost, increased productivity, and the elimination of human error in reading and processing measurements, by utilizing the power of software to perform many of the functions traditionally done by workbenches loaded with hardware.

这样的仪器能够降低劳动成本、提高生产率,减少读取和处理中的人为失误,并可通过使用软件来实现许多传统上由硬件搭建的结构所实现的功能。

# Exercises

### 1. Keywords

In the unit, there are some important words which are the soul of this paper. After reading this paper, we can find some words to stand for this article. Now please find out key words of this paper.

### 2. Summary

After learning the unit, please write a summary about Oscilloscope with 200~300 words.

# 科技英语知识 5：长句的翻译

科技英语中由于大量使用定语从句、状语从句和各种短语，常使句子具有较长的复杂结构，加之英汉两种语言在词法、句法和逻辑思维等方面的差异，造成很多长句难懂难译。

对待长句，要分清层次、突出重点，搞清各分句的内容和结构。译文要逻辑严密，前后呼应，借助语法关系和逻辑承接语，使译文前后衔接、相互呼应，把原文复杂的概念准确通顺地表达出来。

长句的翻译方法通常有顺序法、变序法、分句法、并句法等几种，这些方法可根据不同情况单独或结合使用。

当原句层次分明，和汉语语序相近时可采用顺序法。

例：No such limitation is placed on an AC motor; here the only requirement is relative motion, and since a stationary armature and a rotating field system have numerous advantages, this arrangement is standard practice for all synchronous motor rated above a few kilovolt—amperes.

交流电机不受这种限制，唯一的要求是相对运动，并且由于固定电枢及旋转磁场系统具有很多优点，所以这种安排是所有容量在几千伏·安以上的同步电机的标准做法。

英语习惯前果后因和定状语后置，译成汉语时可采用变序法，即先译后部，再依次向前，改变原句的语序，按汉语习惯的语序译出，从而便于表达和读者理解。

例：The resistance of any length of a conducting wire is easily measured by finding the potential difference in volts between its ends when a known current is following.

已知导线中流过的电流，只要测出导线两端电位差的伏特值，就能很容易得出任何长度导线的电阻值。

为了使译文结构清楚，合乎表达习惯，有时可用拆译法（也称分句法），即将原句分成几个独立的小句，顺序基本不变，保持前后连贯。

例：This kind of two-electrodes tube consists of a tungsten filament, which gives off electrons when it is heated, and a plate toward which the electrons migrate when the field is in the right direction.

这种二极管由一根钨丝和一个极板组成。钨丝受热时放出电子，当电场方向为正时，电子就移向极板。

把两个 which 引导的定语从句从原句中分出，拆成两个独立分句，更符合汉语的表达习惯，意思明确，通俗易懂。

并句法是将原句中的某些部分合并翻译以使译文简洁通顺。

例：It is common practice that electric wires are made from copper.

电线通常是铜制的。

把主句和从句合二为一，避免了冗长复杂的表述方式。

以上几种方法在不同情况下可以灵活掌握，结合使用。下面的例子中采用了拆译和分译相结合的翻译方式。

例：The computer performs a supervisory function in the liquid—level control system by analyzing the process conditions against desired performance criteria and determining the changes in process variables to achieve optimum operation.

在液位控制系统中，计算机实现一种监控功能。它根据给定的特性指标来分析各种过程条件，并决定各过程变量的变化，以实现最佳操纵。

# Unit 7　Linear Circuit Analysis

## Lesson 17　Ohm's Law

Ohm's Law is the relationship of voltage, resistance, and current in dc electrical circuits. It is the most used and certainly the most powerful law in the study of electronics.

### The discovery of Ohm's Law

Ohm's Law, that electric current is proportional to a potential difference, was first discovered by Henry Cavendish, but Cavendish did not publish his discovery. Instead, Ohm published it, and it subsequently came to bear his name. The law appeared in the famous book Die galvanische Kette, mathematisch bearbeitet (The Galvanic Circuit Investigated Mathematically) (1827) in which he gave his complete theory of electricity. The book begins with the mathematical background necessary for an understanding of the rest of the work. While his work greatly influenced the theory and applications of current electricity, it was coldly received at that time. A detailed study of the conceptual framework used by Ohm in formulating Ohm's Law has been presented by Archibald.

### The equation of Ohm's Law

Ohm's Law is an extremely useful equation in the field of electrical/electronic engineering because it describes how voltage, current and resistance are interrelated on a "macroscopic" level, that is, commonly, as circuit elements in an electrical circuit [1].

The mathematical equation that describes this relationship is:

$$I = \frac{V}{R}$$

where $I$ is the current in amperes (often shortened to "amps"), $V$ is the potential difference in volts, and $R$ is a circuit parameter called the resistance (measured in ohms, also equivalent to volts per ampere). The potential difference is also known as the voltage drop, and is sometimes denoted by $U$, $E$ or emf (electromotive force) instead of $V$.

Physicists often use the continuum form of Ohm's Law: $J = \sigma E$, where $J$ is the current density (current per unit area, unlike the simpler $I$, units of amperes, of Ohm's Law), $\sigma$ is the conductivity and $E$ is the electric field (units of volts per meter, unlike the simpler $V$, units of volts, of Ohm's Law).

The potential difference between two points is defined as

$$\Delta V = -\int E dl$$

with the element of path along the integration of electric field vector $E$. For a uniform applied field and defining the voltage in the usual convention of opposite direction to the field: $V=El$.

Substituting current per unit area, $J$, for $I/\alpha$ ($\alpha$ being the cross section of the conductor), the continuum form becomes:

$$\frac{I}{\alpha} = \frac{\sigma V}{l}$$

The electrical resistance of a uniform conductor is given, in terms of conductivity, by:

$$R = \frac{l}{\sigma \alpha}$$

After substitution Ohm's Law takes on the more familiar, yet macroscopic and averaged version:

$$I = \frac{V}{R}$$

### Temperature effects

When the temperature of the conductor increases, the collisions between electrons and ions increase. Thus a substance heats up because of electricity flowing through it (or by any heating process), the resistance will usually increase. The exception is semiconductors. The resistance of an ohmic substance depends on temperature in the following way:

$$R = R_0[\alpha(T - T_0) + 1]$$

where $T$ is its temperature, $T_0$ is a reference temperature (usually room temperature), $R_0$ is the resistance at $T_0$, and $\alpha$ is the percentage change in resistivity per unit temperature. The constant $\alpha$ depends only on the material being considered. The relationship stated is actually only an approximate one, the true physics being somewhat non-linear, or looking at it another way, $\alpha$ itself varies with temperature. For this reason it is usual to specify the temperature that $\alpha$ was measured at with a suffix, such as $\alpha 15$ and the relationship only holds in a range of temperatures around the reference.

### Application to circuit analysis

Resistors in series: A network of interconnected components forms a circuit. In drawing a circuit using a schematic diagram it is assumed that the "wires" (lines) shown connecting the components are perfect conductors relative to the components themselves. When resistors (or other components) are connected sequentially, end to end, they are said to be in series. The total end-to-end resistance of a string of resistors sequentially connected in series is the sum of the resistances, so for the string the total resistance of the string is:

$$R_{total} = R_1 + R_2 + \ldots + R_n \quad \text{(in series)}$$

Resistors in parallel: When resistors are connected side by side they are said to be in parallel. The equation for the total resistance for a group of resistors in parallel is:

$$R_{total} = \frac{1}{\frac{1}{R_1} + \frac{1}{R_2} + \ldots + \frac{1}{R_n}} \quad \text{(in parallel)}$$

Example: A string of resistors in series can be conceptually for analysis purposes replaced by an equivalent resistor having the same total resistance as those series resistors. Similarly, a group of parallel resistors can be conceptually replaced by a single equivalent resistor having the same total resistance as those parallel resistors. This allows Ohm's Law to be applied to a more complicated circuit. Resistor $R_3$ is in series with the parallel combination $R_2//R_1$ (the use of two parallel bars "//" denotes "parallel"). The total resistance across the terminals can be computed by adding the resistance of $R_3$ to the equivalent resistance of the parallel combination $R_2//R_1$:

$$R_{total} = R_3 + R_2//R_1 = R_3 + \frac{1}{\frac{1}{R_2} + \frac{1}{R_1}}$$

If we now apply a voltage $V_{total}$ across the terminals, we can compute the current into the terminals using Ohm's Law:

$$I = \frac{V}{R} = \frac{V_{total}}{R_{total}}$$

Suppose $V_{total} = 12V$, $R_3 = 2\,\Omega, R_2 = 10\Omega$, $R_1 = 15\Omega$. The current into the terminals is found using the above as follows:

$$R_2//R_1 = \frac{1}{\frac{1}{R_2} + \frac{1}{R_1}} = \frac{1}{\frac{1}{10} + \frac{1}{15}} = 6\Omega$$

$$R_{total} = R_3 + R_2//R_1 = 2 + 6 = 8\Omega$$

$$I = \frac{V}{R} = \frac{V_{total}}{R_{total}} = \frac{12V}{8\Omega} = 1.5A$$

The voltage across $R_3$ (denoted $V_{R3}$) could now be calculated from Ohm's Law, since the current through it is known to be 1.5 amperes: $V_{R3} = I_{R3}R_3 = 1.5A \times 2\Omega = 3V$. The voltage across $R_2$ (which is the same as the voltage across $R_1$) could be calculated using Ohm's Law by multiplying the current $I = 1.5A$ times the equivalent parallel resistance $R_2//R_1 = 6\Omega$, giving $1.5 \times 6 = 9V$, or could be calculated by subtracting the voltage across $R_3$ ($V_{R3}$, just calculated above) from the applied voltage of 12V, that is, $12V - 3V = 9V$. Once this is known the current through $R_2$ (denoted $I_{R2}$) may be calculated from Ohm's Law:

$$I_{R2} = \frac{V_{R2}}{R_2} = \frac{9V}{10\Omega} = 0.9A$$

The current through $R_1$ could similarly be found using Ohm's Law by dividing the voltage across it (9V) by its resistance (15$\Omega$), giving 0.6A through $R_1$. Notice that the current through $R_2$ (0.9A) plus the current through $R_1$ (0.6A) equals the total current into the terminals of 1.5A.

### Transients and AC circuits

Ohm's Law holds for linear circuits where the current and voltage are steady (DC), and for the instantaneous voltage and current in linear circuits having only resistive elements and no reactive elements. When reactive elements such as capacitors, inductors, or transmission lines are involved in a circuit to which AC or time-varying voltage or current is applied, the relationship between voltage and current becomes the solution to a differential equation.

Equations for time-invariant AC circuits take the same form as Ohm's Law, however, if the variables are generalized to complex numbers and the current and voltage waveforms are complex exponentials.

In this approach, a voltage or current waveform takes the form $Ae^{st}$, where $t$ is time, $s$ is a complex parameter, and $A$ is a complex scalar. In any linear time-invariant system, all of the currents and voltages can be expressed with the same $s$ parameter as the input to the system, allowing the time-varying complex exponential term to be canceled out and the system described algebraically in terms of the complex scalars in the current and voltage waveforms.

The complex generalization of resistance is impedance, usually denoted $Z$; it can be shown that for an inductor

$$Z = sL$$

and for a capacitor

$$Z = \frac{1}{sC}$$

We can now write

$$V = IZ$$

Where $V$ and $I$ are the complex scalars in the voltage and current respectively and $Z$ is the complex impedance. While this has the form of Ohm's Law, with $Z$ taking the place of $R$, it is not the same as Ohm's Law. When $Z$ is complex, only the real part is responsible for dissipating heat.

In the general AC circuit, $Z$ will vary strongly with the frequency parameter s, and so also will the relationship between voltage and current.

For the common sinusoidal case, the $s$ parameter is taken to be $j\omega$, corresponding to a complex sinusoid $Ae^{j\omega t}$. The real parts of such complex current and voltage waveforms describe the actual sinusoidal currents and voltages in a circuit, which can be in different phases due to the different complex scalars.

# New Words

subsequently ['sʌbsikwəntli] adv. 后来，随后
contiguous [kən'tigjuəs] adj. 邻近的，接近的
continuum [kən'tinjuəm] n. 连续统一体
framework ['freimwəːk] n. 构架，框架，结构
terminal ['təːminl] n. 终点站，终端，接线端
collision [kə'liʒən] n. 碰撞，冲突
exception [ik'sepʃən] n. 例外
instantaneous [ˌinstən'teiniəs] adj. 瞬间的

## Phrases & Expressions

begin with　　首先
take on　　呈现，具有
in terms of　　根据，按照
end-to-end　　端对端
responsible for　　是造成……的原因

## Technical Terms

vector ['vektə] *n.* [数]矢量，向量
potential difference　　位差
current density　　电流密度
reference temperature　　参考温度
real part　　实数

## Note

1. Ohm's Law is an extremely useful equation in the field of electrical/electronic engineering because it describes how voltage, current and resistance are interrelated on a "macroscopic" level, that is, commonly, as circuit elements in an electrical circuit.

欧姆定律是电学与电子工程领域中一个非常有用的方程，因为它描述了"宏观"水平上电压、电流和电阻之间的关系，这里的"宏观"指的是电路中的电子元器件。

这个句子的框架结构是 Ohm's Law is an extremely useful equation。Because 引导原因状语从句，从句中的连接代词 how 引导宾语从句。

## Lesson 18　Kirchhoff's Laws

Kirchhoff's circuit laws are two equalities that deal with the conservation of charge and energy in electrical circuits, and were first described in 1845 by Gustav Kirchhoff. Widely used in electrical engineering, they are also called Kirchhoff's Rules or simply Kirchhoff's Laws.

### Kirchhoff's Current Law (KCL)

This law is also called Kirchhoff's Point Rule, Kirchhoff's Junction Rule (or Nodal Rule), and Kirchhoff's First Rule. The junction between two elements is called a simple node and no division of current results. The junction of three or more elements is called a principal node, and here current division does take place. So the Kirchhoff's Current Law states: The sum of the currents entering a particular point must be zero. Sometime, we also say that the basis for the law is the conservation of electric charge. The principle of conservation of electric charge implies that: At any point in an electrical circuit that does not represent a capacitor plate, the sum of currents flowing towards that point is equal to the sum of currents flowing away from that point. The more simplified form the law is stated as "The algebraic sum of current at a junction is zero". Now let's look at an example such as: we observe four currents "entering" the junction depicted as the bold black dot in Figure 18-1.

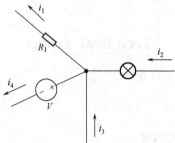

Figure 18-1  The current entering, any junction is equal to the current leaving that junction, $i_1 + i_4 + i_2 + i_3 = 0$

Of course, two currents are actually exiting the junction, but for the purposes of circuit analysis it is generally less restrictive to consider what are in actuality positive currents flowing out of a junction to be negative currents flowing into that junction. Doing so allows us to write Kirchhoff's Law for this example as:

$$\sum i_n = i_1 + i_2 + i_3 + i_4 = 0$$

It may not be clear at this point why we insist on thinking of negative currents flowing into a junction instead of positive currents flowing out. But note that Figure 18-1 provides us with more information than we generally can expect to get when analyzing circuits, namely the helpful arrows indicating the direction of current flow. If we don't have such assistance, we generally should not pass judgment on the direction of current flow (placing a negative sign before our current variable) until we calculate it, lest we confuse ourselves and make mistakes.

Nevertheless in this case we have the extra information of directional arrows in Figure 18-1, so we should take advantage of it. We know that currents $i_2$ and $i_3$ flow into the junction and the currents $i_1$ and $i_4$ flow out. Thus we can write $i_1 + i_4 = i_2 + i_3$. Kirchhoff's Current Law as written is only applicable to steady-state current flow (no alternating current, no signal transmission).

Kirchhoff's Current Law is used in a method of circuit analysis referred to as nodal analysis. A node is a section of a circuit where there is no change in voltage. Each node is used to form an equation, and the equations are then solved simultaneously, giving the voltages at each node.

Example 1:

Consider Figure 18-2 with the following Parameters: $V_1 = 15V$, $V_2 = 7V$, $R_1 = 20\Omega$, $R_2 = 5\Omega$, $R_3 = 10\Omega$. Find current through $R_3$ using Kirchhoff's Current Law.

Solution:

Figure18-3 shows Voltages at Nodes a, b, c and d. We use node a as common node (ground if you like). Thus $V_a = 0V$. From Node b we get: $V_b = -V_1 = -15V$. From Node d we get: $V_d = V_2 = 7V$.

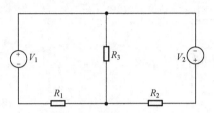 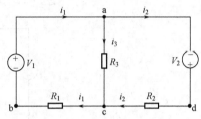

Figure18-2  Example 1    Figure 18-3  Voltages at nodes

It is clear that we must solve $V_c$, in order to complete Voltage definitions at all *nodes*. $V_c$ will be found by applying KCL at Node c and solving resulting equations follows:

$$i_1 = i_2 + i_3$$

$$\frac{V_b - V_c}{R_1} = \frac{V_c}{R_3} + \frac{V_c - V_d}{R_2}$$

and

$$\frac{V_b}{R_1} - \frac{V_c}{R_1} - \frac{V_c}{R_3} - \frac{V_c}{R_2} + \frac{V_d}{R_2} = 0$$

We can group like terms to get the following equation:

$$V_c \left( \frac{1}{R_1} + \frac{1}{R_2} + \frac{1}{R_3} \right) = \frac{V_b}{R_1} + \frac{V_d}{R_2}$$

Substitute values into previous equations you get:

$$V_c \left( \frac{1}{20\Omega} + \frac{1}{5\Omega} + \frac{1}{10\Omega} \right) = \frac{-15V}{20\Omega} + \frac{7V}{5\Omega}$$

$V_c \times 0.35 = 0.65V$, thus $V_c = 1.857V$.

Thus now we can calculate Current through $R_3$ as follows:

$$I_{R3} = \frac{V_c}{R_3} = \frac{1.857}{10} A \approx 0.186 A$$

Just as we expected!

### Kirchhoff's Voltage Law (KVL)

Kirchhoff's Voltage Law states: "The sum of the voltage around a closed circuit path must be zero." Notice that a closed circuit path insists that if one circuit element is chosen as a starting point, then one must be able to traverse the circuit elements in that loop and return to the element in the beginning[1]. Mathematically, the Kirchhoff's Voltage Law is given by:

$$\sum V_n = 0$$

For reference, this law is sometimes called Kirchhoff's Second Law, Kirchhoff's Loop Rule, and Kirchhoff's Second Rule.

We observe five voltages in Figure 18-4: $V_4$ across a voltage source, and the four voltages $V_1, V_2, V_3$ and $V_5$ across the resistors $R_1$, $R_2$, $R_3$ and $R_5$, respectively. The voltage source and resistors $R_1$, $R_2$ and $R_3$ comprise a closed circuit path, thus the sum of the voltages $V_4$, $V_1$, $V_2$ and $V_3$ must be zero:

$$\sum V_n = V_4 + V_3 + V_2 + V_1 = 0$$

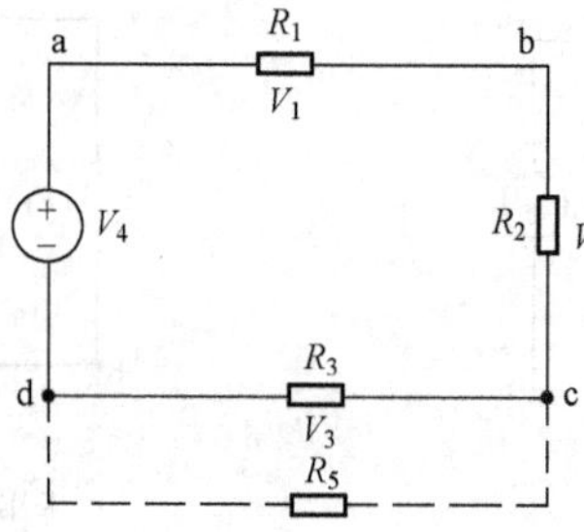

Figure 18-4  $V_1 + V_2 + V_3 + V_4 = 0$

The resistor $R_5$ is outside the closed path in question, and thus plays no role in the calculation of Kirchhoff's Voltage Law for this path. (Note that alternate closed paths can be defined which include the resistor $R_5$. In these cases, the voltage $V_5$ across $R_5$ must be considered in calculating Kirchhoff's Voltage Law.)

Now, if we take the point d in the image as our reference point and arbitrarily set its voltage to zero, we can observe how the voltage changes as we traverse the circuit clockwise.

Going from point d to point a across the voltage source, we experience a voltage increase of $V_4$ volts (as the symbol for the voltage source in the image indicates that point a is at a positive voltage with respect to point d).

On traveling from point a to point b, we cross a resistor. We see clearly from the diagram that, since there is only a single voltage source, current must flow from it's positive terminal to its negative terminal — clockwise around the circuit path. Thus from Ohm's Law, we observe that the voltage drops from point a to point b across resistor $R_1$.

Likewise, the voltage drops across resistors $R_2$ and $R_3$. Having crossed $R_2$ and $R_3$, we arrive back at point d, where our voltage is zero (just as we defined). So we experienced one increase in voltage and three decreases in voltages as we traversed the circuit.

The implication from Kirchhoff's Voltage Law is that, in a simple circuit with only one voltage source and any number of resistors, the voltage drop across the resistors is equal to the voltage applied by the voltage source: $V_4 = V_1 + V_2 + V_3$ [2].

Kirchhoff's Voltage Law can easily be extended to circuitry that contains capacitors.

Example 2:

Consider Figure 18-5 with the following parameters: $V_1 = 15V$, $V_2 = 7V$, $R_1 = 20\Omega$, $R_2 = 5\Omega$, $R_3 = 10\Omega$. Find current through $R_3$ using Kirchhoff's Voltage Law.

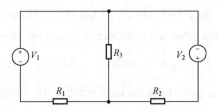

Figure 18-5   Example 2

Solution:

We can see that there are two closed paths (loops) where we can apply KVL in, Loop 1 and 2 as shown in Figure 18-6. From Loop 1 we get: $V_1 - V_{R3} - V_{R1} = 0$. From Loop 2 we get: $V_2 - V_{R3} - V_{R2} = 0$. The above results can further be simplified as follows:

$$V_1 - (I_1 - I_2)R_3 - I_1 R_1 = 0$$

$$I_2 = \frac{I_1(R_3 + R_1) - V_1}{R_3} \tag{18-1}$$

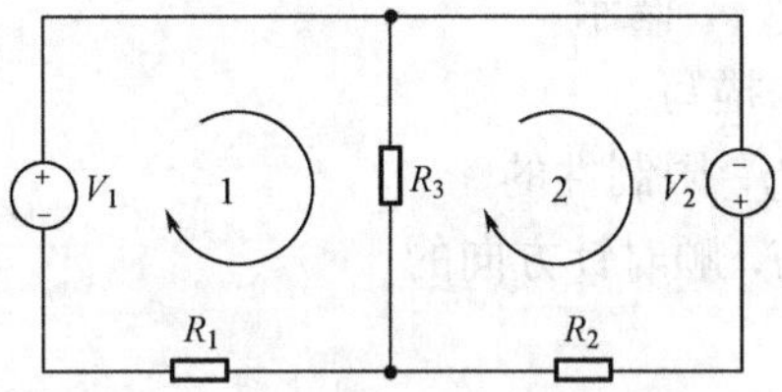

Figure 18-6   Example 2 loops

And $V_2 + (I_1 - I_2)R_3 - I_2 R_2 = 0$.

$$I_2 = \frac{I_1 R_3 + V_2}{R_2 + R_3} \tag{18-2}$$

By equating above (18-1) and (18-2) we can eliminate $I_2$ and hence get the following:

$$I_1 = \frac{V_2 R_3 + V_1(R_2 + R_3)}{(R_2 + R_3)(R_3 + R_1) - R_3^2} \tag{18-3}$$

We end up with the above three equations and now substitute the Values given in the above equations and solve the variables.

If you feel lost up to this point do go back to the beginning of the example. Think of this as just another mathematical problem requiring solving by use of simultaneous equations with two unknowns! Notice that we work with variables only and try to solve the equation to its simplest form. Only after we have arrived at a simplified equation then that we can substitute in values of resistors, voltages and current. This can save you a lot of trouble because if you go wrong you can easily trace your work to the problem.

It is clear that: from (18-3):

$$I_1 = \frac{(7V) \times (10\Omega) + (15V) \times (15\Omega)}{(15\Omega) \times (30\Omega) - (10\Omega)^2} = \frac{295V}{350\Omega} = 0.843A$$

Substitute the above result into (18-2):

$$I_2 = \frac{(7V) + (0.843A \times 10\Omega)}{15\Omega} = \frac{15.43V}{15\Omega} = 1.029A$$

99

The Positive sign for $I_2$ only tells us that Current $I_2$ flows in the same direction to our initial assumed direction. Thus now we can calculate current through $R_3$ as follows:

$$I_{R3} = I_1 - I_2 = 0.843A - 1.029A = -0.186A$$

The Negative sign for $I_{R3}$ only tells us that current $I_{R3}$ flows in the same direction to $I_2$ direction.

## New Words

conservation [ˌkɔnsə(:)'veiʃən] n. 保存，保藏
algebraic [ˌældʒi'breik] adj. 代数的
assistance [ə'sistəns] n. 协助，援助，补助
simultaneously [siməl'teiniəsly] adv. 同时地
implication [ˌimpli'keiʃən] n. 暗示
depict [di'pikt] vt. 描述，描写
restrictive [ris'triktiv] adj. 限制性的
clockwise ['klɔkwaiz] adj. 顺时针方向的

## Phrases & Expressions

insist on　坚持，坚决要求
take advantage of it　利用
return to　恢复，重新采取
play a role of　在……起作用
in the case of　在……的情况

## Technical Terms

node [nəud] n. 节点
parameter [pə'ræmitə] n. 参数
loop [luːp] n. 回路
Kirchhoff's Current Law　基尔霍夫电流定律
Kirchhoff's Voltage Law　基尔霍夫电压定律
principal node　主节点
steady-state current flow　稳态电流
circuit analysis　电路分析
node analysis　节点分析

**Notes**

1. Notice that a closed circuit path insists that if one circuit element is chosen as a starting point, then one must be able to traverse the circuit elements in that loop and return to the element in the beginning.

注意,闭合通路指的是,如果选择电路中的一个元器件作为起点,那必须能够穿过回路中所有的元器件并且回到初始元器件。

这个句子的框架结构是 notice ... 一个祈使句。第一个 that 引导主句的宾语从句;从句的 that 同样引导宾语从句,在第二个宾语从句中,if 引导条件状语从句。then one must be able to 中的 one 指代前面的 one circuit element。

2. The implication from Kirchhoff's Voltage Law is that, in a simple circuit with only one voltage source and any number of resistors, the voltage drop across the resistors is equal to the voltage applied by the voltage source: $V_4 = V_1 + V_2 + V_3$。

由基尔霍夫电压定律的含义可知,在只有电压源和许多电阻的简单电路中,通过电阻的电压降等于压源提供的电压,即 $V_4 = V_1 + V_2 + V_3$。

这个句子的框架结构是 The implication is that ... , that 引导表语从句,从句中主语是 the voltage drop,谓语是 is equal to,宾语是 the voltage applied by the voltage source: $V_4 = V_1 + V_2 + V_3$。其中, in a simple circuit with only one voltage source and any number of resistors 是状语。

# Lesson 19  Circuit Analysis Methods

In this section Kirchhoff's Current Law (KCL) and Kirchhoff's Voltage Law (KVL) will be used to determine currents and voltages throughout a network. For simplicity, we will first illustrate the basic principles of both node analysis and mesh analysis using only DC circuits. Once the fundamental concepts have been explained and illustrated, we will demonstrate the generality of both analysis techniques through an AC circuit example.

Node analysis

In a node analysis, the node voltages are the variables in a circuit, and KCL is the vehicle used to determine them. One node in the network is selected as a reference node, and then all other node voltages are defined with respect to that particular node. This reference node is typically referred to as ground using the symbol ($\equiv$), indicating that it is at ground-zero potential. Consider the network shown in Figure 19-1. The network has three nodes, and the nodes at the bottom of the circuit has been selected as the reference node. Therefore the two remaining nodes, labeled $V_1$ and $V_2$, are measured with respect to this reference node. Suppose that the node voltages $V_1$ and $V_2$ have somehow been determined, i. e., $V_1 = 4V$ and

$V_2 = -4$V. Once these node voltages are known, Ohm's Law can be used to find all branch currents. For example,

$$I_1 = \frac{V_1 - 0}{2} = 2\text{A}$$

$$I_2 = \frac{V_1 - V_2}{2} = \frac{4 - (-4)}{2} = 4\text{A}$$

$$I_3 = \frac{V_2 - 0}{1} = \frac{-4}{1} = -4\text{A}$$

Note that KCL is satisfied at every node, i. e.,

$$I_1 - 6 + I_2 = 0$$
$$-I_2 + 8 + I_3 = 0$$
$$-I_1 + 6 - 8 - I_3 = 0$$

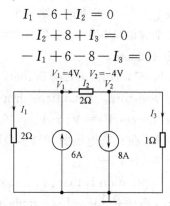

Figure 19-1  A three-node network

Therefore, as a general rule, if the node voltages are known, all branch currents in the network can be immediately determined. In order to determine the node voltages in a network, we apply KCL to every node in the network except the reference node. Therefore, given an $N$-node circuit, we employ $N-1$ linearly independent simultaneous equations to determine the $N-1$ unknown node voltages. Graph theory, can be used to prove that exactly $N-1$ linearly independent KCL equations are required to find the $N-1$ unknown node voltages in a network. Let us now demonstrate the use of KCL in determining the node voltages in a network. For the network shown in Figure 19-2, the bottom node is selected as the reference and the three remaining nodes, labeled $V_1$, $V_2$, and $V_3$, are measured with respect to that node. All unknown branch currents are also labeled. The KCL equations for the three nonreference nodes are

$$I_1 + 4 + I_2 = 0$$
$$-4 + I_3 + I_4 = 0$$
$$-I_1 - I_4 - 2 = 0$$

Using Ohm's Law these equations can be expressed as:

$$\frac{V_1 - V_3}{2} + 4 + \frac{V_1}{2} = 0$$

$$-4 + \frac{V_2}{1} + \frac{V_2 - V_3}{1} = 0$$

$$-\frac{(V_1 - V_3)}{2} - \frac{(V_2 - V_3)}{1} - 2 = 0$$

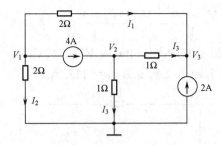

Figure 19-2    A four-node network

Solving these equations, using any convenient method, yields $V_1=-8/3\text{V}, V_2=10/3\text{V}$, and $V_3=8/3\text{V}$. Applying Ohm's Law we find that the branch currents are $I_1=-16/6\text{A}$, $I_2=-8/6\text{A}, I_3=20/6\text{A}$, and $I_4=4/6\text{A}$. A quick check indicates that KCL is satisfied at every node. The circuits examined thus far have contained only current sources and resistors. In order to expand our capabilities, we next examine a circuit containing voltage sources. The circuit shown in Figure 19-3 has three nonreference nodes labeled $V_1$, $V_2$, and $V_3$. However, we do not have three unknown node voltages. Since known voltage sources exist between the reference node and nodes $V_1$ and $V_3$, these two node voltages are known, i. e., $V_1=12\text{V}$ and $V_3=-4\text{V}$. Therefore, we have only one unknown node voltage, $V_2$.

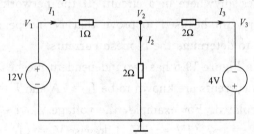

Figure 19-3    A four-node network containing voltage sources

The equations for this network are then $V_1=12\text{V}, V_3=-4\text{V}$ and $-I_1+I_2+I_3=0$.
The KCL equation for node $V_2$ written using Ohm's Law is

$$-\frac{(12-V_2)}{1}+\frac{V_2}{2}+\frac{V_2-(-4)}{2}=0$$

Solving this equation yields $V_2=5\text{V}$, $I_1=7\text{A}$, $I_2=5/2\text{A}$, and $I_3=9/2\text{A}$. Therefore, KCL is satisfied at every node.

Thus, the presence of a voltage source in the network actually simplifies a node analysis. In an attempt to generalize this idea, consider the network in Figure 19-4.

Note that in this case $V_1=12\text{V}$ and the difference between node voltages $V_3$ and $V_2$ is constrained to be 6V. Hence, two of the three equations needed to solve for the node voltages in the network are $V_1=12\text{V}, V_3-V_2=6\text{V}$.

To obtain the third required equation, we form what is called a supernode, indicated by the dotted enclosure in the network. Just as KCL must be satisfied at any node in the network, it must be satisfied at the supernode as well. Therefore, summing all the currents leaving the supernode yields the equation:

$$\frac{V_2 - V_1}{1} + \frac{V_2}{2} + \frac{V_3 - V_1}{1} + \frac{V_3}{2} = 0$$

The three equations yield the node voltages $V_1 = 12V$, $V_2 = 5V$, and $V_3 = 11V$, and therefore $I_1 = 1A$, $I_2 = 7A$, $I_3 = 5/2A$, and $I_4 = 11/2A$.

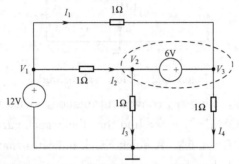

Figure 19-4  A four-node network used to illustrate a supernode

**Mesh analysis**

In a mesh analysis the mesh currents in the network are the variables and KVL is the mechanism used to determine them. Once all the mesh currents have been determined, Ohm's Law will yield the voltages anywhere in a circuit. If the network contains $N$ independent meshes, then graph theory can be used to prove that $N$ independent linear simultaneous equations will be required to determine the $N$ mesh currents.

The network shown in Figure 19-5 has two independent meshes. They are labeled $I_1$ and $I_2$, as shown. If the mesh currents are known to be $I_1 = 7A$ and $I_2 = 5/2A$, then all voltages in the network can be calculated. For example, the voltage $V_1$, i. e. , the voltage across the $1\Omega$ resistor, is $V_1 = -I_1 R = -7 \times 1V = -7V$. Likewise $V_2 = (I_1 - I_2)R = (7 - 5/2) \times 2V = 9V$. Furthermore, we can check our analysis by showing that KVL is satisfied around every mesh. Starting at the lower left-hand corner and applying KVL to the left-hand mesh we obtain $-7 \times 1 + 16 - (7 - 5/2) \times 2 = 0$ where we have assumed that increases in energy level are positive and decreases in energy level are negative. Consider now the network in Figure 19-6 Once again, if we assume that an increase in energy level is positive and a decrease in energy level is negative, the three KVL equations for the three meshes defined are $-I_1(1) - 6 - (I_1 - I_2)(1) = 0$, $+12 - (I_2 - I_1)(1) - (I_2 - I_3)(2) = 0$, $-(I_3 - I_2)(2) + 6 - I_3(2) = 0$, $2I_1 - I_2 = -6$, $-I_1 + 3I_2 - 2I_3 = 12$, $-2I_2 + 4I_3 = 6$.

Solving these equations using any convenient method yields $I_1 = 1$ A, $I_2 = 8$ A, and $I_3 = 11/2$ A. Any voltage in the network can now be easily calculated, e. g. , $V_2 = (I_2 - I_3)(2) = 5V$ and $V_3 = I_3(2) = 11V$. Just as in the node analysis discussion, we now expand our capabilities by considering circuits which contain current sources. In this case, we will show that for mesh analysis, the presence of current sources makes the solution easier. The network in Figure 19-7 has four meshes which are labeled $I_1$, $I_2$, $I_3$, and $I_4$. However, since two of these currents, i. e. , $I_3$ and $I_4$, pass directly through a current source, two of the four linearly independent equations required to solve the network are $I_3 = 4, I_4 = -2$.

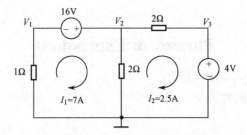

Figure 19-5  A network containing two independent meshes

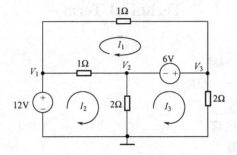

Figure 19-6  A three-mesh network

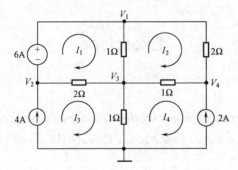

Figure 19-7  A four-mesh network

The two remaining KVL equations for the meshes defined by $I_1$ and $I_2$ are $+6-(I_1-I_2)(1)-(I_1-I_3)(2)=0, -(I_2-I_1)(1)-I_2(2)-(I_2-I_4)(1)=0$.

Solving these equations for $I_1$ and $I_2$ yields $I_1 = 54/11$ A and $I_2 = 8/11$ A. A quick check will show that KCL is satisfied at every node. Furthermore, we can calculate any node voltage in the network. For example, $V_3 = (I_3-I_4)(1) = 6$ V and $V_1 = V_3 + (I_1-I_2)(1) = 112/11$ V.

# New Words

simplicity [sim'plisiti] *n.* 简单,简易,朴素,直率
demonstrate ['demənstreit] *vt.* 示范,证明,论证
generalize ['dʒenərəlaiz] *vt.* 归纳,概括,推广
ground [graund] *n.* 接地
label ['leibl] *n.* 标签

## Phrases & Expressions

with respect to   关于,(至于)谈到
be referred to   被提及,被提交
in order to   为了……

## Technical Terms

mesh [meʃ] *n.* 网格
network ['netwəːk] *n.* 网络
potential [pə'tenʃ(ə)l] *n.* 电位
general rule   通用定理
branch current   支路电流
current source   电流源

# Exercises

### 1. Keywords

In article, there are some important words which are the soul of this paper. After reading this paper, we can find some words to stand for this article. Now please find out key words of this paper.

### 2. Summary

Both node analysis and mesh analysis are the most important method of linear circuit analysis. Do you have some idea with this article? Please write a summary about this paper with less 200 words.

# 科技英语知识6：专业英语词汇特点

**词汇的专业性含义强**

科技文章中涉及大量的专业术语，这些词汇有的仅是某专业的专用词，不具有其他含义，如 diode(二极管)、capacitor(电容)等；而有些则除了具有本专业含义外，还在其他专业和日常生活中具有不同的含义，如 spring(春天、喷泉、弹簧)、bus(公共汽车、总线)、memory(记忆、内存)等。后一类词需要在平时进行积累，并且根据不同的语言环境和专业进行适当的翻译，否则将会引起误解。

**缩略词多**

缩略词或缩写词经常出现在科技文章中，通常有两种情况，一是一些专业领域常用的缩略词，二是在某篇文章中作者针对该文使用的术语转换成的缩略词，此类缩略词首次出现时都会在文章中进行标注。在平时的阅读时要注意积累本专业常见的缩略词，对科技文献的阅读是大有裨益的。例如：

AC(alternate current) 交流电；
DC(direct current) 直流电；
IC(integrated circuit) 集成电路；
CPU(central processor unit) 中央处理器；
FET(field effect transistor) 场效应管；
EMF(electromotive force) 电动势。

**复合词和派生词多**

科技文章的一大特点就是大量使用复合词和派生词。随着科技发展的需要，经常会组合或派生出一些新的词汇，有些词可能无法在字典中查到，对这类词汇要通过分析了解其含义。

**1. 复合词**

例如：
open-loop 开环；
close-loop 闭环；
feedback 反馈；
bandwidth 带宽。

**2. 派生词**

派生词通常的构成为：前缀＋词根＋后缀。

例如：semi＋conduct＋or＝semiconductor
semi(半)＋conduct(传导)＋or(名词后缀)＝半导体 $n.$。

因此，在平时多注意积累前后缀和词根，能有效地扩充词汇量。

例如：
electro-(电，电的)；
mocro-(微小的)；
therm-(热的)；

# Unit 8　Digital Logic Circuit

## Lesson 20　Digital Systems

　　Digital systems are designed to store, process, and communicate information in digital form. They are found in a wide range of applications, including process control, communication systems, digital instruments, and consumer products. The digital computer, more commonly called the "computer", is an example of a typical digital system.

　　A computer manipulates information in digital, or more precisely, binary form. A binary number has only two discrete values—zero or one. Each of these discrete values is represented by the OFF and ON status of an electronic switch called a "transistor". All computers, therefore, only understand binary numbers. Any decimal number (base 10, with ten digits from 0 to 9) can be represented by a binary number (base 2, with digits 0 and 1). The basic blocks of a computer are the central processing unit (CPU), the memory, and the input/output (I/O). The CPU of the computer is basically the same as the brains of a human being. Computer memory is conceptually similar to human memory. A question asked to a human being is analogous to entering a program into the computer using an input device such as the keyboard, and answering the question by the human is similar in concept to outputting the result required by the program to a computer output device such as the printer[1]. The main difference is that human beings can think independently, whereas computers can only answer questions that they are programmed for. Programs can perform a specific task such as addition if the computer has an electronic circuit capable of adding two numbers. Programmers cannot change these electronic circuits but can perform tasks on them using instructions.

　　Computer software, on the other hand, consists of a collection of programs. Programs contain instructions and data for performing a specific task. These programs, written using any programming language such as C++, must be translated into binary prior to execution by the computer. This is because the computer only understands binary numbers. Therefore, a translator for converting such a program into binary is necessary.

　　Hence, a translator program called the compiler is used for translating programs written in a programming language such as C++ into binary. These programs in binary form are then stored in the computer memory for execution because computers only understand 1's and 0's. Furthermore, computers can only add. This means that all operations such as subtraction, multiplication, and division are performed by addition.

　　Due to advances in semiconductor technology, it is possible to fabricate the CPU in a single chip. The result is the microprocessor. Both Metal Oxide Semiconductor (MOS) and

Bipolar technologies were used in the fabrication process. The CPU can be placed on a single chip when MOS technology is used. However, several chips are required with the bipolar technology. HCMOS (High Speed Complementary MOS) or BICMOS (Combination of Bipolar and HCMOS) technology is normally used these days to fabricate the microprocessor in a single chip. Along with the microprocessor chip, appropriate memory and I/O chips can be used to design a microcomputer. The pins on each one of these chips can be connected to the proper lines on the system bus, which consists of address, data, and control lines. In the past, some manufacturers have designed a complete microcomputer on a single chip with limited capabilities. Single-chip microcomputers were used in a wide range of industrial and home applications.

"Microcontrollers" evolved from single-chip microcomputers. The microcontrollers are typically used for dedicated applications such as automotive systems, home appliances, and home entertainment systems. Typical microcontrollers, therefore, include a microcomputer, timers, and A/D (analog to digital) and D/A (digital to analog) converters all in a single chip. Examples of typical microcontrollers are Intel 8751 (8-bit) / 8096 (16-bit) and Motorola HC11 (8-bit) / HC16 (16-bit).

## New Words

manipulate [mə'nipjuleit] *vt.* 操纵,利用,操作
discrete [dis'kri:t] *adj.* 不连续的,离散的;[计] 离散的
decimal ['desiməl] *adj.* 十进位的,小数的;*n.* 十进制,小数;[计] 十进制,小数点
conceptually [kən'septʃuəli] *adv.* 概念地
bipolar [bai'pəulə] *adj.* [电]两极的,双极的;偶极的

## Phrases & Expressions

be analogous to　类似于,与……相似
on the other hand　另一方面
translate into　翻译成,转化为

## Technical Terms

translator [træns'leitə] *n.* 翻译者;[计] 翻译程序,翻译器
compiler [kəm'pailə] *n.* [计](自动编码器)编译程序器;程序编制(器)
semiconductor technology　[计]半导体工艺,半导体技术
microprocessor　[计]微处理器;微处理机

**Note**

1. A question asked to a human being is analogous to entering a program into the computer using an input device such as the keyboard, and answering the question by the human is similar in concept to outputting the result required by the program to a computer output device such as the printer.

以键盘等输入装置输入程序就相当于人类提出问题,而解决问题就是由计算机通过输出设备,如打印机将程序所得的结果输出。

## Lesson 21  Logic Gates

Digital circuits contain hardware elements called "gates" that perform logic operations on binary numbers. Devices such as transistors can be used to perform the logic operations. Boolean algebra is a mathematical system that provides the basis for these logic operations. George Boole, an English mathematician, introduced this theory of digital logic. The term *Boolean* **variable** is used to mean the two-valued binary digit 1 or 0.

Boolean algebra uses three basic logic operations namely, NOT, OR, and AND. These operations are described next.

### NOT operation

The NOT operation inverts or provides the ones complement of a binary digit. This operation takes a single input and generates one output. The NOT operation of a binary digit provides the following result:

$$\text{NOT } 1 = 0$$
$$\text{NOT } 0 = 1$$

Therefore, NOT of a Boolean variable $A$, written as $\overline{A}$ (or $A'$) is 1 if and only if $A$ is 0. Similarly, $\overline{A}$ is 0 if and only if $A$ is 1. This definition may also be specified in the form of a truth table:

| Input | Output |
|---|---|
| $\overline{A}$ | $A$ |
| 0 | 1 |
| 1 | 0 |

Note that a truth table contains the inputs and outputs of digital logic circuits. The symbolic representation of an electronic circuit that implements a NOT operation is shown in Figure 21-1.

A NOT gate is also referred to as an "inverter" because it inverts the voltage levels. A transistor acts as an inverter. A 0-volt at the input generates a 5-volt output; a 5-volt input provides a 0-volt output. As an example, the 74HC04 (or 74LS04) is a hex inverter 14-pin chip containing six independent inverters in the same chip as shown in

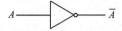

Figure 21-1   NOT gate

Figure 21-2. Computers normally include a NOT instruction to perform the ones complement of a binary number on a bit-by-bit basis. An 8-bit computer can perform NOT operation on an 8-bit binary number. For example, the computer can execute a NOT instruction on an 8-bit binary number 01101111 to provide the result 10010000. The computer utilizes an internal electronic circuit consisting of eight inverters to invert the 8-bit data in parallel.

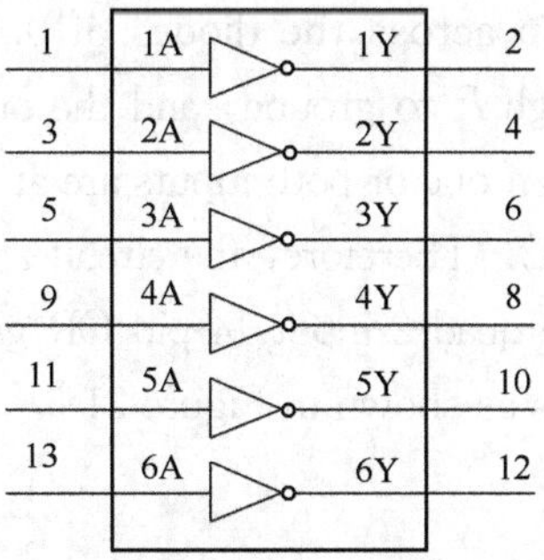

Figure 21-2   Pin diagram for the 74HC04 or 74LS04

## OR operation

The OR operation for two variables $A$ and $B$ generates a result of 1 if $A$ or $B$, or both, are 1. However, if both $A$ and $B$ are zero, then the result is 0.

A plus sign $+$ (logical sum) or $\vee$ symbol is normally used to represent OR. The four possible combinations of ORing two binary digits are:

$$0 + 0 = 0$$
$$0 + 1 = 1$$
$$1 + 0 = 1$$
$$1 + 1 = 1$$

A truth table is usually used with logic operations to represent all possible combinations of inputs and the corresponding outputs. The truth table for the OR operation is:

| Input<br>A   B | Output<br>A+B |
|---|---|
| 0   0 | 0 |
| 0   1 | 1 |
| 1   0 | 1 |
| 1   1 | 1 |

Figure 21-3 shows the symbolic representation of an OR gate.

Figure 21-3  OR gate

Logic gates using diodes provide good examples to understand how semiconductor devices are utilized in logic operations. Note that diodes are hardly used in designing logic gates. Figure 21-4 shows a two-input-diode OR gate. The diode is a switch, and it closes when there is a voltage drop of 0.6V between the anode and the cathode. Suppose that a voltage range of 0 to 2V is considered as logic 0 and a voltage of 3 to 5V is logic 1. If both $A$ and $B$ are at logic 0 (say 1.5V) with a voltage drop across the diodes of 0.6V to close the diode switches, a current flows from the inputs through $R$ to ground, and the output $C$ will be at $1.5V - 0.6V = 0.9V$ (logic 0). On the other hand, if one or both inputs are at logic 1 (say 4.5V) the output $C$ will be at $4.5V - 0.6V = 3.9V$ (logic 1). Therefore, the circuit acts as an OR gate. The 74HC32 (or 74LS32) is a commercially available quad 2-input 14-pin OR gate chip. This chip contains four 2-input/1-output independent OR gates as shown in Figure 21-5.

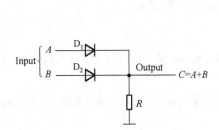

Figure 21-4  Diode OR gate

Figure 21-5  Pin diagram for 74HC32 or 74LS32

### AND operation

The AND operation for two variables $A$ and $B$ generates a result of 1 if both $A$ and $B$ are 1.

However, if either $A$ or $B$, or both, are zero, then the result is 0. The dot • and $\wedge$ symbol are both used to represent the AND operation. The AND operation between two binary digits is

$$0 \cdot 0 = 0$$
$$0 \cdot 1 = 0$$
$$1 \cdot 0 = 0$$
$$1 \cdot 1 = 1$$

The truth table for the AND operation is:

| Input | Output |
|-------|--------|
| A   B | A · B  |

| Input | Output |
|-------|--------|
| 0 0   | 0      |
| 0 1   | 0      |
| 1 0   | 0      |
| 1 1   | 1      |

Figure 21-6 shows the symbolic representation of an AND gate. Figure 21-7 shows a two-input diode AND gate.

Figure 21-6 AND gate symbol

As we did for the OR gate, let us assume that the range 0 to $+2V$ represents logic 0 and the range 3 to 5V is logic 1. Now, if $A$ and $B$ are both HIGH (say 3.3V) and the anode of both diodes at 3.9V, the switches in $D_1$, and $D_2$, close. A current flows from $+5V$ through resistor $R$ to $+3.3V$ input to ground. The output $C$ will be HIGH (3.9V). On the other hand, if a low voltage (say 0.5V) is applied at $A$ and a high voltage (3.3V) is applied at $B$. The value of $R$ is selected in such a way that 1.1V appears at the anode side of $D_1$, at the same time 3.9V appears at the anode side of $D_2$. A current flows from the $+5V$ input through $R$ and the diodes to ground. Output $C$ will be low (1.1V) because the output will be lower of the two voltages. Thus, it can be shown that when either one or both inputs are low, the output is low, so the circuit works as an AND gate. As mentioned before, diode logic gates are easier to understand, but they are not normally used these days.

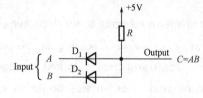

Figure 21-7  Diode AND gate

The 74HC08 (or 74LS08) is a commercially available quad 2-input 14-pin AND gate chip. This chip contains four 2-input/1-output independent AND gates as shown in Figure 21-8.

Figure 21-8  Pin diagram for 74HC08 or 74LS08

## New Words

inverter [in'və:tə] n. 倒转者，换流器，反相器；[计] 反相器
diode ['daiəud] n. 二极管
anode ['ænəud] n. [电] 阳极，正极，板极，屏板，氧化极
cathode ['kæθəud] n. [电] 阴极，负极
quad [kwɔd] n. [计] 四芯导线

## Phrases & Expressions

refer to　　参考，查阅；涉及，提到；指的是

## Technical Terms

OR　　或操作，或运算
AND　　与操作，与运算
Boolean algebra　　[计] 逻辑代数；布尔代数

## Lesson 22　Flip-flop

In digital circuits, a flip-flop is a term referring to an electronic circuit (a bistable multivibrator) that has two stable states and thereby is capable of serving as one bit of memory. Today, the term flip-flop has come to mostly denote non-transparent (clocked or edge-triggered) devices, while the simpler transparent ones are often referred to as latches, however, as this distinction is quite new, the two words are sometimes used interchangeably[1].

A flip-flop is usually controlled by one or two control signals and/or a gate or clock signal. The output often includes the complement as well as the normal output. As flip-flops are implemented electronically, they require power and ground connections.

The first electronic flip-flop was invented in 1918 by William Eccles and F. W. Jordan. It was initially called the Eccles-Jordan trigger circuit and consisted of two active elements. The name flip-flop was later derived from the sound produced on a speaker connected to one of the back coupled amplifiers outputs during the trigger process within the circuit. This original electronic flip-flop — a simple two-input bistable circuit without any dedicated clock signal, was transparent, and thus a device that would be labeled as a "latch" in many circles today.

Flip-flops can be either simple or clocked. Simple flip-flops can be built around a pair of cross-coupled inverting elements: vacuum tubes, bipolar transistors, field effect transistors, inverters, and inverting logic gates have all been used in practical circuits-perhaps augmented by some gating mechanism [2]. The more advanced clocked devices are specially designed for synchronous systems; such devices therefore ignore its inputs except at the transition of a dedicated clock signal. This causes the flip-flop to either change or retain its output signal based upon the values of the input signals at the transition. Some flip-flops change output on the rising edge of the clock, others on the falling edge.

Clocked flip-flops are typically implemented as master-slave devices where two basic flip-flops (plus some additional logic) collaborate to make it insensitive to spikes and noise between the short clock transitions; they nevertheless also often include asynchronous clear or set inputs which may be used to change the current output independent of the clock [3].

Flip-flops can be further divided into types that have found common applicability in both asynchronous and clocked sequential systems: the SR (set-reset), D (data or delay), T (trigger), and JK types are the common ones; all of which may be synthesized from other types by a few logic gates [4]. The behavior of a particular type can be described by what is termed the characteristic equation, which derives the "next" (after the next clock pulse) output, $Q_{next}$, in terms of the input signal(s) and/or the current output, Q.

### Set-reset flip-flops (SR flip-flops)

The fundamental latch is the simple SR flip-flop, where S and R stand for set and reset respectively. It can be constructed from a pair of cross-coupled NAND or NOR logic gates. The stored bit is present on the output marked Q.

Normally, in storage mode, the S and R inputs are both low, and feedback maintains the Q and $\overline{Q}$ outputs in a constant state, with $\overline{Q}$ the complement of Q. If S is pulsed high while R is held low, then the Q output is forced high, and stays high even after S returns low; similarly, if R is pulsed high while S is held low, then the Q output is forced low, and stays low even after R returns low. Figure 22-1 shows the symbol for a SR latch.

| SR flip-flop operation | | | | | | | |
|---|---|---|---|---|---|---|---|
| Characteristic table | | | Excitation table | | | | |
| S | R | Action | $Q(t)$ | $Q(t+1)$ | S | R | Action |
| 0 | 0 | Keep state | 0 | 0 | 0 | X | No change |
| 0 | 1 | Q=0 | 0 | 1 | 1 | 0 | Set |
| 1 | 0 | Q=1 | 1 | 0 | 0 | 1 | Reset |
| 1 | 1 | Unstable combination | 1 | 1 | X | 0 | No change |

("X" denotes a Don't care condition; meaning the signal is irrelevant)

Figure 22-1  The symbol for a SR latch

### Trigger flip-flops (T flip-flops)

If the T input is high, the T flip-flop changes state ("toggles") whenever the clock input is strobed. If the T input is low, the flip-flop holds the previous value. This behavior is described by the characteristic equation:

$Q_{next} = T \oplus Q$ (or, without benefit of the XOR operator, the equivalent: $Q_{next} = \overline{T}Q \oplus T\overline{Q}$) and can be described in a truth table:

| T flip-flop operation | | | | | | | |
|---|---|---|---|---|---|---|---|
| Characteristic table | | | | Excitation table | | | |
| T | Q | $Q_{next}$ | Comment | Q | $Q_{next}$ | T | Comment |
| 0 | 0 | 0 | Hold state | 0 | 0 | 0 | No change |
| 0 | 1 | 1 | Hold state | 1 | 1 | 0 | No change |
| 1 | 0 | 1 | Toggle | 0 | 1 | 1 | Complement |
| 1 | 1 | 0 | Toggle | 1 | 0 | 1 | Complement |

Figure 22-2 shows a T flip-flop. When T is held high, the trigger flip-flop divides the clock frequency by two; that is, if clock frequency is 4 MHz, the output frequency obtained from the flip-flop will be 2 MHz. This "divide by" feature has application in various types of digital counters. A T flip-flop can also be built using a JK flip-flop (J & K pins are connected together and act as T) or D flip-flop (T input and $Q_{previous}$ is connected to the D input through an XOR gate).

### JK flip-flop

The JK flip-flop augments the behavior of the SR flip-flop (J = Set, K = Reset) by interpreting the S = R = 1 condition as a "flip" or toggle command. Specifically, the combination J = 1, K = 0 is a command to set the flip-flop; the combination J = 0, K = 1 is a command to reset the flip-flop; and the combination J = K = 1 is a command to toggle the flip flop, i. e., change its output to the logical complement of its current value. Setting J = K = 0 does NOT result in a D flip-flop, but rather, will hold the current state. To synthesize a D flip-flop, simply set K equal to the complement of J. The JK flip-flop is therefore a universal flip-flop, because it can be configured to work as an SR flip-flop, a D flip-flop, or a T flip-flop.

Figure 22-2  T flip-flop

NOTE: The flip-flop is positive edge triggered (Clock Pulse) as seen in Figure 22-3.

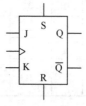

Figure 22-3  JK flip-flop

The characteristic equation of the JK flip-flop is:
$$Q_{next} = J\overline{Q} + \overline{K}Q$$
and the corresponding truth table is:

| JK flip-flop operation | | | | | | | | |
|---|---|---|---|---|---|---|---|---|
| Characteristic table | | | | Excitation table | | | | |
| J | K | $Q_{next}$ | Comment | Q | $Q_{next}$ | J | K | Comment |
| 0 | 0 | $Q_{prev}$ | Hold state | 0 | 0 | 0 | X | No change |
| 0 | 1 | 0 | Reset | 1 | 1 | 1 | X | Set |
| 1 | 0 | 1 | Set | 1 | 0 | X | 1 | Reset |
| 1 | 1 | $\overline{Q}_{prev}$ | Toggle | 1 | 1 | X | 0 | No change |

### D flip-flop

The Q output always takes on the state of the D input at the moment of a rising clock edge (or falling edge if the clock input is active low). It is called the D flip-flop for this reason, since the output takes the value of the D input or Data input, and Delays it by one clock count. The D flip-flop can be interpreted as a primitive memory cell, zero-order hold, or delay line. Figure 22-4 shows D flip-flop.

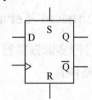

Figure 22-4  D flip-flop

Truth table:

| Clock | D | Q | $Q_{prev}$ |
|---|---|---|---|
| Rising edge | 0 | 0 | x |
| Rising edge | 1 | 1 | x |
| No rising | x | $Q_{prev}$ | |

("x" denotes a Don't care condition, meaning the signal is irrelevant)

These flip flops are very useful, as they form the basis for shift registers, which are an essential part of many electronic devices. The advantage of the D flip-flop over the D-type latch is that it "captures" the signal at the moment the clock goes high, and subsequent changes of the data line do not influence Q until the next rising clock edge. An exception is that some flip-flops have a "reset" signal input, which will reset Q (to zero), and may be either asynchronous or synchronous with the clock.

## New Words

bistable [bai'steibl] *adj.* 双稳态的,双稳定的;[计]双稳态的
transparent [træns'pɛərənt] *adj.* 透明的,显然的,清晰的;[计]透明
interchangeably [ˌintəːˈtʃeindʒəˈbli] *adv.* 可交换地,可交替地,可互换地
asynchronous [eiˈsiŋkrənəs] *adj.* 不同时的,异步的;[计]异步的
configure [kənˈfiə] *vt.* 装配,使成形;[计]配置
subsequent [ˈsʌbsikwənt] *adj.* 后来的,并发的;其次的,接着发生的

## Phrases & Expressions

derive from    得自,由来,衍生
divide into    分开;划分;分成
cross-coupled    交叉耦合
i. e.    即,也就是

## Technical Terms

multivibrator [ˈmʌltivaiˈbreitə] *n.* [计]多谐振荡器
field effect transistor    [计]场效应晶体管
shift register    移位寄存器

## Notes

1. Today, the term flip-flop has come to mostly denote non-transparent (clocked or edge-triggered) devices, while the simpler transparent ones are often referred to as latches; however, as this distinction is quite new, the two words are sometimes used interchangeably.
今天,触发器用来表示大部分非透明(时钟触发或边沿触发)器件,而简单透明的触发器往往指的是锁存器;但是,由于这一区别用法提出时间不长,所以这两个词有时可以互换使用。

2. Simple flip-flops can be built around a pair of cross-coupled inverting elements: vacuum tubes, bipolar transistors, field effect transistors, inverters, and inverting logic gates have all been used in practical circuits—perhaps augmented by some gating mechanism.
单一触发器可以由一对交叉耦合反相器件搭建,真空管、双极晶体管、场效应晶体管、反相器和反相逻辑门都已应用在实际电路中,这可能增加了一些门控机制。

3. Clocked flip-flops are typically implemented as master-slave devices where two basic flip-flops (plus some additional logic) collaborate to make it insensitive to spikes and noise between the short clock transitions; they nevertheless also often include asynchronous clear or set inputs which may be used to change the current output independent of the clock.

时钟触发器通常由主从方式实现,主从触发由两个基本触发器(加上一些外围逻辑)级联组成,这种结构在时钟间隔时间短时对毛刺和噪声不敏感。此外,还可以不依赖时钟进行异步清零和置位来改变输出。

4. Flip-flops can be further divided into types that have found common applicability in both asynchronous and clocked sequential systems: the SR (set-reset), D (data or delay), T (trigger), and JK types are the common ones; all of which may be synthesized from other types by a few logic gates.

触发器可以进一步划分成不同的类型,这些类型已经应用在异步系统和时序系统中,常见的类型是SR(置位-复位)、D(数据或延迟)、T(双态)和JK触发器。所有这些触发器都可以由其他类型触发器通过添加少许逻辑门来合成。

# Exercises

1. Keywords

In the unit, there are some important words which are the soul of this paper. After reading each paper, we can find some words to stand for the article. Now please find out key words(3~5)of this paper.

2. Summary

After learning the unit, please write a summary about Logic gate with 200~300 words.

# 科技英语知识7：专业英语概述

根据国家教委要求，大学生在经过基础英语的学习后，基本上已掌握了英语的常用语法和4000以上的词汇量，具备较扎实的英语基础。进入三年级后，随着专业课的进一步学习，学生的专业知识技能也开始逐步加强。具备以上两个条件后，应进行专业英语的训练，在保证20万字以上阅读量的基础上，对本专业英文资料的阅读应达到基本的要求。因此，掌握专业英语技能是大学基础英语学习的主要目的之一，是一种素质的提高，直接关系到学生的求职和毕业后的工作能力。

专业英语的重要性体现在很多方面，大到日益广泛的国际间的科学技术交流，小到产品说明书的翻译，以及作为网络主要语言的英语对工程技术人员提供的巨大的专业信息量对资料查询者提出了更高的要求。

尽管很多人在此之前已经进行了多年的基础英语学习，但专业英语的学习仍是很必要的。

首先，专业英语在词义上具有不同于基础英语的特点和含义。例如：

If a mouse is installed in a computer, then the available memory space for user will reduce.

错误译法：如果让老鼠在计算机里筑窝，那么使用者的记忆空间就会减少。

专业译法：如果计算机安装了鼠标，则用户可以利用的内存空间就会减少。

Connect the black pigtail with the dog-house.

错误译法：把黑色的猪尾巴系在狗窝上。

专业译法：将黑色的引出线接在高频高压电源屏蔽罩上。

通过以上的例子，我们不难认识到专业词汇的一些特点。同一个词在日常生活与不同的专业领域中可能会有不同的含义，单纯依靠日常用语进行望文生义的解释不仅会闹笑话，还有可能出事故。

其次，外文科技文章在结构上也具有很多自身的特点，如长句多、被动语态多、大量的名词化结构等，都给对原文的理解和翻译带来了日常基础英语所难以解决的困难。

再次，专业英语对听、说、读、写、译的侧重点不同，最主要的要求在于"读"和"译"，也就是通过大量的阅读对英文资料进行正确的理解和翻译，在读和译的基础上，对听、说、写进行必要的训练。

最后，专业英文资料由于涉及很多科技内容而往往极为复杂，难以理解，加上这类文章的篇幅通常很长，所以只有经过一定的专业英语训练，才能完成从基础英语到专业英语的过渡，达到学以致用的最终目的。

专业翻译是指把科技文章由原作语言（source language）用译文语言（target language）忠实、准确、严谨、通顺、完整地再现出来的一种语言活动。它要求翻译者在具有一定专业基础知识和英语技能的前提下，借助合适的英汉科技词典来完成整个翻译过程。专业翻译直接应用于科技和工程，因而对翻译的质量有极高的要求。翻译上的失之毫厘，工程中就有可能差之千里，造成巨大的损失。

例如，有这样一个标志牌：

Control Center. Smoking Free.

它的意思是"控制中心，严禁吸烟"，free 在这里作"免除……的"讲，而如果理解为"随便的，自由的"，就会产生完全相反的效果。

# Unit 9  Integrated Circuits

## Lesson 23  Wafers

A wafer is a thin slice of semiconductor material, such as a silicon crystal, used in the fabrication of integrated circuit and other microdevices.

The wafer serves as the substrate for microelectronic devices built in and over the wafer and undergoes many microfabrication process steps such as doping or ion implantation, etching, deposition of various materials, and photolithographic patterning.

Wafer Fabrication is a procedure composed of many repeated sequential processes to produce complete electrical or photonic circuits. Examples of these include production of radio frequency (RF) amplifiers, LEDs, optical computer components, and CPUs for computers. Wafer fabrication is used to build components with the necessary electrical structures.

The main process begins with electrical engineers designing the circuit and defining its functions, and specifying the signals, inputs, outputs and voltages needed[1]. These electrical circuit specifications are entered into electrical circuit design software, such as SPICE, and then imported into circuit layout programs. This is necessary for the layers to be defined for wafer mask production. An etched silicon wafer is shown in Figure 23-1.

Wafers are formed of highly pure (99.9999% purity), nearly defect-free single crystalline material. Wafer thickness is determined by the mechanical strength of the material used. The wafer must be thick enough to support its own weight without cracking during handling.

Figure 23-1  An etched silicon wafer

The silicon wafers start out blank and pure. But silicon wafers are generally not 100% pure silicon. The circuits are built in layers in clean rooms. The silicon wafer shaping involves a series of precise mechanical and chemical process steps. Wafers are grown from crystal having a regular crystal structure, with silicon having a diamond cubic structure with a lattice spacing of 5.430710Å (0.5430710nm). When cut into wafers, the surface is aligned in one of several relative directions known as crystal orientations.

Orientation is defined by the Miller index with <100> or <111> faces being the most common for silicon. Orientation is important since many properties of a single crystal's structural and electronic are highly anisotropic. Ion implantation depths depend on the wafer's crystal orientation, since each direction offers distinct paths for transport.

Wafer cleavage typically occurs only in a few well-defined directions. Scoring the wafer along cleavage planes allows it to be easily diced into individual chips so that the billions of individual circuit elements on an average wafer can be separated into many individual circuits. Silicon wafers are available in a variety of sizes from 25.4mm (1inch) to 300mm (11.8 inches). Semiconductor fabrication plants (also known as fabs) are defined by the size of wafers that they are tooled to produce.

The size has gradually increased to improve throughput and reduce cost with the current state-of-the-art fab considered to be 300mm (12inches), with the next standard set to be 450mm (18inches) [2]. Intel, TSMC and Samsung are separately conducting research to the advent of 450mm "prototype" (research) fabs by 2012. Figure 23-2 shows silicon wafer.

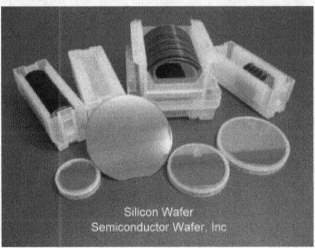

Figure 23-2  Silicon wafer

1 inch.
2 inches (50.8mm). Thickness 275μm.
3 inches (76.2mm). Thickness 375μm.
4 inches (100mm). Thickness 525μm.
5 inches (127mm) or 125 mm (4.9 inches). Thickness 625μm.
5.9 inches (150mm, usually referred to as "6 inches"). Thickness 675μm.

7.9 inches (200mm, usually referred to as "8 inches"). Thickness 725μm.

11.8 inches (300mm, usually referred to as "12 inches" or "Pizza size" wafer). Thickness 775μm.

18 inches (450mm). Thickness 925μm (expected).

## New Words

substrate ['sʌbstreit] n. (＝substratum)衬底；[地]底土层,本源；[生]培养基；[生化]酶作用物
undergo [ˌʌndə'gəu] vt. 经历,遭受,忍受
etch [etʃ] v. 刻蚀
deposition [ˌdepə'ziʃən, di:-] n. 沉积作用,沉积物,革职,废王位,免职
procedure [prə'si:dʒə] n. 程序,手续
optical ['ɔptikəl] adj. 眼的,视力的,光学的
crystalline ['kristəlain] adj. 水晶的
crack [kræk] n. 裂缝,噼啪声；v. (使)破裂,裂纹,(使)爆裂；adj. 最好的,高明的
align [ə'lain] vi. 排列；vt. 使结盟,使成一行
anisotropic [əˌnaisəu'trɔpik] adj. 各向异性的
cleavage ['kli:vidʒ] n. 劈开,分裂
prototype ['prəutətaip] n. 原型

## Phrases & Expressions

built in    安装,固定
start out    出发,动身
separate into    分成
state-of-the-art    艺术级的

## Technical Terms

silicon wafer    硅晶圆
diamond cubic structure    金刚石立方结构
crystal orientation    晶向
Miller index    密勒指数
ion implantation    离子注入
wafer fabrication    晶圆制造
radio frequency amplifiers    音频放大器
mechanical strength    机械强度

## Notes

1. The main process begins with electrical engineers designing the circuit and defining its functions, and specifying the signals, inputs, outputs and voltages needed.

晶圆制造的主要过程始于电子工程师的电路设计,定义电路的功能,描述信号、输入、输出及电压。

2. The size has gradually increased to improve throughput and reduce cost with the current state-of-the-art fab considered to be 300 mm (12 inch), with the next standard set to be 450 mm (18 inch).

晶圆的尺寸增长可以提高生产率以及降低成本,当今晶片的主流是 300 毫米(12 英寸),接下来能达到 450 毫米(18 英寸)。

# Lesson 24  IC Processing Technology

In this lesson we describe the common processing techniques for integrated circuits. In order to understand the process, it is necessary to understand these steps. The process steps described here include crystal growth, oxidation, diffusion and ion implantation, lithography and etching.

### Crystal growth

All processing starts with single-crystal silicon material. One process for forming crystalline wafers is known as Czochralski method growth invented by the Polish chemist Jan Czochralski. In this process(Figure 24-1), a seed of crystalline silicon is immersed in molten silicon and gradually pulled out while rotating. As a result, a large single-crystal cylindrical "ingot" is formed that can be sliced thin into wafers. The diameter of the wafer has scaled up with new technology generations, exceed 20cm today.

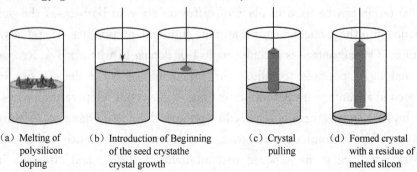

(a) Melting of polysilicon doping  (b) Introduction of Beginning of the seed crystathe crystal growth  (c) Crystal pulling  (d) Formed crystal with a residue of melted silcon

Figure 24-1  The Czochralski process

The crystals are normally grown in either a<100>or <111>crystal orientation. The resulting crystals are cylindrical and have a diameter of 75~300mm and a length of 1m. The cylindrical crystals are sliced into wafers that are approximately 0.5~0.7mm thick for wafers or size 100~150mm, respectively.

### Oxidation

The first basic processing step is oxide growth or oxidation. Oxidation is the process by which a layer of silicon dioxide ($SiO_2$) is formed on the surface of the silicon wafer. Semiconductor can be oxidized by various methods. Among the methods of Oxidation, thermal oxidation is the key process in modern silicon integrated circuits technology.

Silicon dioxide is "grown" by placing the exposed silicon in an oxidizing atmosphere such as oxygen at a temperature around 1000℃. The rate of growth depends on the type and pressure of the atmosphere, the temperature, and the doping level of the silicon[1].

### Diffusion and ion implantation

Diffusion and ion implantation are the two key processes we use to introduce controlled amounts of dopant into semiconductor. Both diffusion and ion implantation are used fabricating discrete devices and integrated circuit. They are used to dope selectively the semiconductor substrate to produce either an n-or p-type region.

Ion implantation is widely used in the fabrication of MOS components. Ion implantation is the process by which ions of a particular dopant are accelerated by an electric field to high velocity and physically lodge within the semiconductor material. The average depth of penetration varies from 0.1 to 0.6μm depending on the velocity and angle at which the ions strike the silicon wafer. The path of each ion depends on the collisions it experiences Therefore, ions are typically implanted off-axis from the wafer so that they will experiences collisions with lattice atoms, thus avoiding undesirable channeling of ions deep into the silicon.

An alternative method to address channeling is to implant through silicon dioxide, which randomizes the implant direction before the ions enter the silicon. The ion-implant process causes damage to the semiconductor crystal lattice, leaving many of the implanted ions electrically inactive. This damage can be repaired by an annealing process.

Ion implantation can be used in place of diffusion since in both cases the objective is to insert impurities into the semiconductor material. Ion implantation has several advantages over thermal diffusion. One advantage is accurate control of doping-to within±5%. Reproducibility is very good, making it possible to adjust the thresholds of MOS devices or create precise resistor. A second advantage is that ion implantation is a room temperature process, although annealing at higher temperatures is required to remove the crystal damage, A third advantage is that it is possible to implant through a thin layer. Consequently, the material to be implanted does not have to be exposed to contaminants during and after the implantation process. Unlike ion implantation, diffusion requires that the surface be free of silicon dioxide or silicon nitride layers. Finally, ion implantation allows controller over the profile of the implanted impurities. For example concentration peak can be placed below the surface of the silicon if desired.

## Lithographies

Each of the basic semiconductor fabrication processes discussed thus far is only applied to selected parts of the silicon wafer with the exception of oxidation and deposition. The selections of these parts are accomplished by a process called lithographies.

There are several lithographic methods to implement. For example, optical lithography, electron beam lithography, X-ray lithography, and ion beam lithography. However, each method has its limitation, diffraction effects in optical lithography, proximity effects in electron beam lithography and the disadvantages are high cost and low throughput mask fabrication complexities in X-ray lithography, and beam deflection difficulties in ion beam lithography.

Let us consider the n-well pattern as a example. This pattern is "written" to a transparent glass "mask" by a precisely controlled electron beam [Figure 24-2(a)]. Also, as depicted in Figure 24-2(b), the wafer is covered by a thin layer "photoresist", a material whose etching properties change upon exposure to light, subsequently. The mask is placed on top of the wafer and the pattern is projected onto the wafer by ultraviolent (UV) light [Figure 24-2(c)].

The photoresist "hardens" in the regions exposed to light and remains "soft" under the opaque rectangle, the wafers is then placed in an etchant that dissolves the "soft" photoresist area, thereby exposing the silicon surface [Figure24-2(d)]. Now, and an n-well can be created in the exposed area. We call this set of operations a lithography sequence.

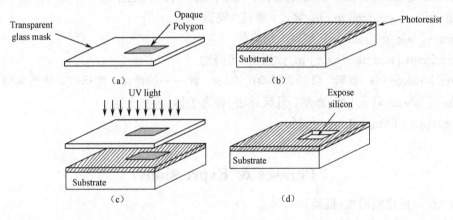

Figure 24-2  The process of the lithographies

## Etching

Etching is the process of removing exposed (unprotected) material. Material that normally etched includes polysilicon, silicon dioxide, silicon nitride, and aluminum.

There are two basic types or etching techniques. Wet etching is used extensively in semiconductor processing.

Wet etching uses chemicals to remove the material to be etched. HF is used to etch silicon dioxide; $H_3PO_4$ is used to remove silicon nitride; nitric acid, acetic acid, or hydrofluoric acid is used to remove polysilicon potassium hydroxide is used to etch silicon; and a phosphoric acid mixture is used to remove metal. The wet-etching technique is strongly dependent on time and temperature, and care must be taken with the acids used in wet etching as the represent a potential hazard. Dry etching or plasma etching uses ionized gases that are rendered chemically active by RF-generated plasma.

This process requires significant characterization to optimize pressure, gas flow rate, gas mixture, and RF power. Dry etching is very similar to sputtering and in fact the same equipment can be used. Reactive ion etching (RIE) induces plasma etching accompanied by ionic bombardment[2]. Dry etching is used for submicron technologies since it achieves anisotropic profiles (no undercutting).

## New Words

oxidation [ɔksi'deiʃən] n. [化]氧化
diffusion [di'fjuːʒən] n. 扩散,传播,漫射
molten ['məultən] v. 熔化; adj. 熔铸的
immerse [i'məːs] vt. 沉浸,使陷入
ingot ['iŋgət] n. [冶]锭铁,工业纯铁
cylindrical [si'lindrik(ə)l] [计]圆柱的
velocity [vi'lɔsiti] n. 速度,速率,迅速,周转率
randomize ['rændəmaiz] v. 使随机化,完全打乱,(使)任意排列
penetration [peni'treiʃən] n. 穿过,渗透,突破
accurate ['ækjurit] adj. 正确的,精确的
contaminant [kən'tæminənt] n. 致污物,污染物
hazard ['hæzəd] n. 冒险,危险,冒险的事; vt. 冒……的危险,赌运气,使遭危险
plasma ['plæzmə] n. [解]血浆,乳浆;[物]等离子体,等离子区
lithography [li'θɔgrəfi] n. 光刻

## Phrases & Expressions

scale up    按比例增加(提高)
slice into    切一个口子
on the surface    表面上
with the exception of    除……以外

## Technical Terms

thermal diffusion    热扩散
concentration peak    浓度峰值
wet etching    湿法刻蚀
dry etching    干法刻蚀

gas flow rate　气流率
crystal growth　单晶生长
Czochralski method　切克劳斯基法(直拉法)
thermal oxidation　热氧化
electric field　电场

## Notes

1. The rate of growth depends on the type and pressure of the atmosphere, the temperature, and the doping level of the silicon.
氧化的生长速度取决于氧化气氛的类型、压强、温度及硅片的掺杂浓度。

2. Dry etching is very similar to sputtering and in fact the same equipment can be used.
干法刻蚀技术十分类似于溅射,而且实际上可以用同种设备来实现。离子反应刻蚀是一种伴有离子轰击的等离子体刻蚀。

# Lesson 25　IC Design Flow

In this lesson, we will introduce design flow of Analog IC and Digital IC. Analog IC design consists of Front-end Design and Back-end Design. The Front-end Design is shown if Figure 25-1. Main steps of the Front-end Design are as follows:

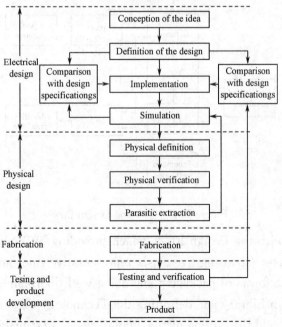

Figure 25-1　Front-end Design flow

1. Knowing well the processing of Foundry, understand of the single MOSFET's $I$-$V$ Curve and reading the simulation files used in .lib;

2. Determining the smallest $W$ and $L$ of the MOSFET, the digital circuits and the witch often taking the smallest $W$ and $L$;

3. Making sure SPEC and knowing well the chip's performance indicators. For example, electrical characteristics and typical performance characteristics;

4. Structuring system diagram and achieve the main function module;

5. Design of the Sub-circuits contain function, structure, performance indicators;

6. The simulations of the Sub-circuits contain functional verification, parameter simulation etc;

7. The simulation of the whole circuit contain functional verification, parameter simulation etc. ;

8. Writing design note and simulationon report, doing well the datas of the front-end design flow;

9. Finishing the whole circuit of analog IC.

The Back-end Design flow (Figure 25-2) is as follows:

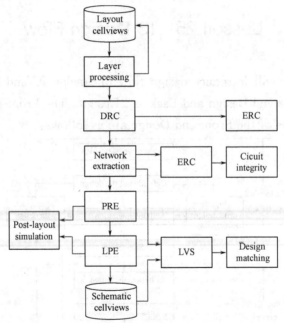

Figure 25-2  Back-end Design flow

1. Being familiar with the Design Rules which providing by Foundry and understanding each design rules of mask. Choosing five or six representative rules and remember them;

2. Making sure the forms of packaging and doing well floorplan of the chip;

3. Establishing own Library and definiting the Technology File and Display File;

4. Reasonable to build the outline of the whole layout, first making sure the location of the pad and the large devices;

5. The layout design of the sub-circuit modules. At the same time, DRC (Design Rule Check) and LVS (Layout Versus Schematic);

6. Completing the overall layout;

7. Executing DRC and LVS;

8. LPE (Layout Parasitic Extraction) of the whole layout, carrying out post simulation;

9. Exporting the GDS File;

10. Delivering Foundry tape out.

The Typical Digital IC Design flow is (Figure 25-3) as follows:

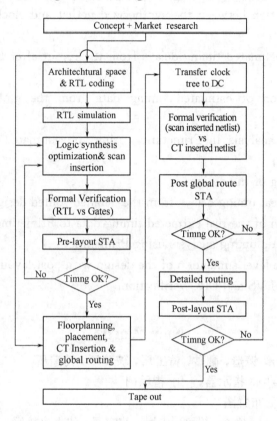

Figure 25-3  The Typical Digital IC Design flow

1. Architectural and electrical specification;

2. RTL coding in HDL;

3. DFT memory BIST insertion, for designs containing memory elements;

4. Exhaustive dynamic simulation of the design, in order to verify the functionality of the design;

5. Design environment setting. This includes the technology library to be used, along with other environmental attributes;

6. Constraining and synthesizing the design with scan insertion (and optional JTAG) using Design Compiler;

7. Block level static timing analysis, using Design Compiler's built-in static timing analysis engine;

8. Formal verification of the design. RTL compared against the synthesized netlist, using Formality;

9. Pre-layout static timing analysis on the full design through Primetime;

10. Forward annotation of timing constraints to the layout tool;

11. Initial floorplanning with timing driven placement of cells, clock tree insertion and global routing [1];

12. Transfer of clock tree to the original design (netlist) residing in Design Compiler;

13. In-place optimization of the design in Design Compiler;

14. Formal verification between the synthesized netlist and clock tree inserted netlist, using Form;

15. Extraction of estimated timing delays from the layout after the global Routing step (step 11);

16. Back annotation of estimated timing data from the global routed design, to Primetime;

17. Static timing analysis in Primetime, using the estimated delays extracted after performing global route;

18. Detailed routing of the design;

19. Extraction of real timing delays from the detailed routed design;

20. Back annotation of the real extracted timing data to Primetime;

21. Post-layout static timing analysis using Primetime;

22. Functional gate-level simulation of the design with post-layout timing (if desired);

23. Tape out after LVS and DRC verification.

## New Words

foundry ['faundri] n. 铸造，翻砂，铸工厂，玻璃厂，铸造厂
indicator ['indikeitə] n. 指示器；[化]指示剂
floorplan n. 平(楼)面布置图
establish [is'tæbliʃ] vt. 建立，设立，安置，使定居，使人民接受，确定；vi. 建立
connectivity [kənek'tiviti] n. 连通性
optimism ['ɔptimizəm] n. 乐观，乐观主义
deflection [di'flekʃən] n. 偏斜，偏转，偏差

## Phrases & Expressions

be familiar with　　熟悉

## Technical Terms

design rule check　　设计规则检查
layout versus schematic　　版图和电路图对照

functional verification　　功能验证
parameter simulation　　参数仿真
layout parasitic extraction　　版图参数提取

## Note

1. Initial floorplanning with timing driven placement of cells, clock tree insertion and global routing.

具有时序驱动单元布局、时钟树插入和全局布线的初始布局划分。

## Exercises

### 1. Keywords

In the three lessons, there are some important words which are the soul of the content. After reading the three lessons, we can find some words to stand for each lesson. Now please find out key words of every lesson respectively.

### 2. Summary

After reading this unit, please write a summary of 200～300 words about integrated circuit.

# 科技英语知识 8：数词的翻译

在科技文献中，当利用数词来表示各种数量概念及这些概念的表示方法时，若不确切理解其含义，就可能会产生错译。

**绝对量的译法**

在科技英语中常见用阿拉伯数字表示的数量，翻译时数字不太大（如五六位数以下）的，包括温度、压力、年份、产量、耗量和金额等，一般可照抄，例如：

At 300℃        在 300℃；
by 1760        到 1760 年；
280km/h        280 千米/小时。

较大的数字可以加以换算，用"万"、"亿"等计数单位加以表示，例如：

500MW          50 万千瓦；
70kV           7 万伏；
20000tons      2 万吨。

用文字表示的数量，如果是确定的，可根据具体情况译成数字或中文，如 five hundred thousand 500000（五十万）；对于大约的数量，可以按照语言习惯翻译，如 tens of ...（数十……）、dozens of ...（几十……）、hundreds of ...（几百……）等。

**增减量的译法**

增加量和减少量的表示及译法主要如下。

1. 净增（减）量：所增（减）数字照译。

Both the width and the length are increased by five times.
宽度和长度均增加五倍。
The loss of power was reduced by 40%。
功率损耗减少了 40%。

2. 成 $n$ 倍地增加（减少）可译为"增加（减少）到 $n(1/n)$"或"增加（减少）$n-1$ 倍$((n-1)/n)$"。例如：

By 1975 the production of transistor radios would be expected to increase five-fold.
到 1975 年，晶体管收音机的产量预计将增加 5 倍。

3. "as+形容词或副词"+again as（或将 again 提到最前面）表示"比……大（长、宽……）一倍"（即净增一倍）；如果 again 前面再加 half，则表示"比……大（长、宽……）半倍"。例如：

This wire is as long again as that.
这根金属线的长度为那根的两倍。

4. 用 double、treble、quadruple 等作动词，可表示增加到的倍数。

5. 减少一半。例如：

decrease one-half       （减少一半）；
reducing ... one half   （减少一半）；

cut ... in half　（把……减少一半）；
shorten ... two times　（缩短一半）；
one-half less　（少一半，小一半）；
be less than half　（少一半还多）。
6. 减余量，即减少后剩余的数量，通常用介词 to 和数词 $n$ 来表示。例如：
decrease to 50　（减少到 50）；
reduce to 60%　（减少到 60%）。

# Unit 10  Microcomputer

## Lesson 26  Universal Serial Bus

### What is USB?

Anyone who has been around computers for more than two or three years knows the problem that the Universal Serial Bus is trying to solve-in the past, connecting devices to computers has been a real headache!

- Printers connected to parallel printer ports, and most computers only came with one. Things like Zip drives, which need a high-speed connection into the computer, would use the parallel port as well, often with limited success and not much speed.
- Modems used the serial port, but so did some printers and a variety of odd things like Palm Pilots and digital cameras. Most computers have at most two serial ports, and they are very slow in most cases.
- Devices that needed faster connections came with their own cards, which had to fit in a card slot inside the computer's case. Unfortunately, the number of card slots is limited and you needed a Ph. D, to install the software for some of the cards.

The goal of USB is to end all of these headaches. The Universal Serial Bus gives you a single, standardized, easy-to-use way to connect up to 127 devices to a computer. Connecting a USB device to a computer is simple-you find the USB connector on the back or front of your machine and plug the USB connector into it.

If it is a new device, the operating system auto-detects it and asks for the driver disk. If the device has already been installed, the computer activates it and starts talking to it. USB devices can be connected and disconnected at any time.

A USB cable has two wires for power ($+5$ volts and ground) and a twisted pair of wires to carry the data. Low-power devices(such as mice)can draw their power directly from the bus. High-power devices(such as printers)have their own power supplies and draw minimal power from the bus. Individual USB cables can run as long as 5 meters; with hubs, devices can be up to 30 meters(six cables' worth)away from the host.

Many USB devices come with their own built-in cable, and the cable has an "A" connection on it. If not, then the device has a socket on it that accepts a USB "B" connector. The USB standard uses "A" and "B" connectors to avoid confusion:

- "A" connectors head "upstream" toward the computer.
- "B" connectors head "downstream" and connect to individual devices.

By using different connectors on the upstream and downstream end, it is impossible to ever get confused if you connect any USB cable's "B" connector into a device, you know that it will work. Similarly, you can plug any "A" connector into any "A" socket and know that it will work.

## USB 2.0[1]

The standard for USB version 2.0 was released in April 2000 and serves as an upgrade for USB 1.1[2]. USB 2.0 (High-speed USB) provides additional bandwidth for multimedia and storage applications and has a data transmission speed 40 times faster than USB 1.1. To allow a smooth transition for both consumers and manufacturers, USB 2.0 has full forward and backward compatibility with original USB devices and works with cables and connectors made for original USB, too.

Supporting three speed modes(1.5, 12 and 480 megabits per second), USB 2.0 supports low-bandwidth devices such as keyboards and mice, as well as high-bandwidth ones like high-resolution Webcams, scanners, printers and high-capacity storage systems. The deployment of USB 2.0 has allowed PC industry leaders to forge ahead with the development of next—generation PC peripherals to complement existing high-performance PCs. The transmission speed of USB 2.0 also facilitates the development of next-generation PCs and applications. In addition to improving functionality and encouraging innovation, USB 2.0 increases the productivity of user applications and allows the user to run multiple PC applications at once or several high—performance peripherals simultaneously.

### Data transfer

When the host powers up, it queries all of the devices connected to the bus and assigns each one an address. This process is called enumeration — devices are also enumerated when they connect to the bus. The host also finds out from each device what type of data transfer it wishes to perform:

- Interrupt—a device like a mouse or a keyboard, which will be sending very little data, would choose the interrupt mode.
- Bulk—a device like a printer, which receives data in one big packet, uses the bulk transfer mode. a block of data is sent to the printer(in 64-byte chunks) and verified to make sure it is correct.
- Isochronous—a streaming device(such as speakers) uses the isochronous mode. Data streams between the device and the host in real-time, and there is no error correction.

The host can also send commands or query parameters with control packets.

# New Words

standardize ['stændədaiz] *vt.* 使标准化;用标准校检
built-in [,bilt'in] *adj.* 嵌入的;内置的;固有的
upstream [,ʌp'striːm] *adj.* 向上游的;逆流而上的;(石油工业等)上游的。*adv.* 向上游; 逆流地

downstream [ˌdaun'striːm] adj. 顺流而下的，在下游方向的；adv. 在下游地，顺流地
upgrade [ˌʌp'greid] vt. 提升；使（机器、计算机系统等）升级
innovation [ˌinə'veiʃən] n. 改革，创新；新观念；新发明；新设施
enumeration [iˌnjuːmə'reiʃən] n. 计数，列举；细目；详表；点查
isochronous [ai'sɔkrənəs] adj. 同步的，等时的

## Phrases & Expressions

solve-in 解决
parallel port 并口
Universal Serial Bus 通用串行总线
Palm Pilots 一种掌上平板电脑
Zip 压缩（文档）
auto-detects 自动检测

## Notes

1. USB 2.0：USB 2.0 的最大传输速率高达 480Mbps。
2. USB 1.0：USB 1.0/1.1 的最大传输速率为 12Mbps，1996 年推出。

# Lesson 27  MCS-51

The Intel 8051 microcontroller is one of the most popular general purpose microcontrollers in use today. The success of the Intel 8051 spawned a number of clones which are collectively referred to as the MCS-51 family of microcontrollers, which includes chips from vendors such as Atmel, Philips, Infineon, and Texas Instruments.

The Intel 8051 is an 8-bit microcontroller which means that most available operations are limited to 8 bits. There are 3 basic "sizes" of the 8051：Short, Standard, and Extended. The Short and Standard chips are often available in DIP form, but the Extended 8051 models often have a different form factor, and are not "drop-in compatible". All these things are called 8051 because they can all be programmed using 8051 assembly language, and they all share certain features (although the different models all have their own special features).

Some of the features that have made the 8051 popular are：
- 8-bit data bus;
- 16-bit address bus;
- 32 general purpose registers each of 8 bits;
- 16-bit timers (usually 2, but may have more, or less);

- 3 internal and 2 external interrupts;
- Bit as well as byte addressable RAM area of 16 bytes;
- Four 8-bit ports, (short models have two 8-bit ports);
- 16-bit program counter and data pointer.

8051 models may also have a number of special, model-specific features, such as UARTs, ADC, OpAmps, etc.

### Basic pins

Pin 9: Pin 9 is the reset pin which is used reset the microcontroller's internal registers and ports upon starting up.

Pins 18 and 19: The 8051 has a built-in oscillator amplifier hence we need to only connect a crystal at these pins to provide clock pulses to the circuit.

Pins 40 and 20: Pins 40 and 20 are VCC and ground respectively. The 8051 chip needs + 5V 500mA to function properly, although there are lower powered versions like the Atmel 2051 which is a scaled down version of the 8051 which runs on +3V.

Pins 29, 30 and 31: As described in the features of the 8051, this chip contains a built-in flash memory. In order to program this we need to supply a voltage of +12V at Pin 31. If external memory is connected then Pin 31, also called EA/VPP, should be connected to ground to indicate the presence of external memory. Pin 30 is called ALE (address latch enable), which is used when multiple memory chips are connected to the controller and only one of them needs to be selected. Pin 29 is called PSEN. This is "program select enable". In order to use the external memory it is required to provide the low voltage (0) on both PSEN and EA pins.

### Ports

There are 4 8-bit ports: P0, P1, P2 and P3.

Port P1 (pins 1 to 8): The Port P1 is a general purpose input/output port which can be used for a variety of interfacing tasks. The other ports P0, P2 and P3 have dual roles or additional functions associated with them based upon the context of their usage.

Port P3 (pins 10 to 17): Port P3 acts as a normal I/O port, but Port P3 has additional functions such as, serial transmit and receive pins, 2 external interrupt pins, 2 external counter inputs, read and write pins for memory access.

Port P2 (pins 21 to 28): Port P2 can also be used as a general purpose 8 bits port when no external memory is present, but if external memory access is required then Port P2 will act as an address bus in conjunction with Port P0 to access external memory. Port P2 acts as A8-A15, as can be seen from Figure 27-1.

Port P0 (pins 32 to 39) Port P0 can be used as a general purpose 8 bits port when no external memory is present, but if external memory access is required then Port P0 acts as a multiplexed address and data bus that can be used to access external memory in conjunction with Port P2. P0 acts as AD0~AD7, as can be seen from Figure 27-1.

### Data and program memory

The 8051 Microprocessor can be programmed in PL/M, 8051 Assembly, C and a number of other high-level languages. Many compilers even have support for compiling C++ for an 8051. Program memory in the 8051 is read-only, while the data memory is considered to be read/write accessible. When stored on EEPROM or Flash, the program memory can be rewritten when the microcontroller is in the special programmer circuit.

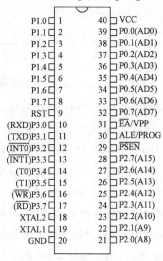

Figure 27-1  Pin diagram of the 8051 DIP

### Program start address

The 8051 starts executing program instructions from address 0x00 in the program.

### Direct memory

The 8051 has 256 bytes of internal addressable RAM, although only the first 128 bytes are available for general use by the programmer. The first 128 bytes of RAM (from 0x00 to 0x7F) are called the direct memory, and can be used to store data.

### Special function register

The special function register (SFR) is the upper area of addressable memory, from address 0x80 to 0xFF. This area of memory cannot be used for data or program storage, but is instead a series of memory-mapped ports and registers. All port input and output can therefore be performed by memory MOV operations on specified addresses in the SFR. Also, different status registers are mapped into the SFR, for use in checking the status of the 8051, and changing some operational parameters of the 8051.

### General purpose registers

The 8051 has 4 selectable banks of 8 addressable 8-bit registers, R0 to R7. This means that there are essentially 32 available general purpose registers, although only 8 (one bank) can be directly accessed at a time. To access the other banks, we need to change the current bank number in the flag status register.

## A and B registers

The A register is located in the SFR at memory location 0xE0. The A register works in a similar fashion to the AX register of x86 processors. The A register is called the accumulator, and by default it receives the result of all arithmetic operations. The B register is used in a similar manner, except that it can receive the extended answers from the multiply and divide operations. When not being used for multiplication and Division, the B register is available as an extra general-purpose register.

Figure 27-2 show internal schematics of the 8051.

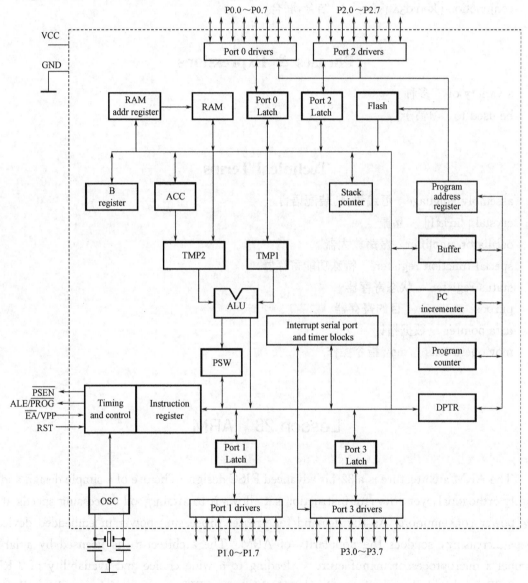

Figure 27-2  Internal schematics of the 8051

## New Words

microcontroller [ˈmaikrəukənˈtrəulə] *n.* 微型控制器
spawn [spɔːn] *n.* (鱼)卵,产物; *vt.* 产卵,酿成,大量生产
clone [kləun] *n.* 无性系;[计]代用件
dual [ˈdjuːəl] *a.* 双重的,双的
multiplication [ˌmʌltipliˈkeiʃən] *n.* 增多,增加,乘,繁殖
conjunction [kənˈdʒʌŋkʃən] *n.* 结合,联合

## Phrases & Expressions

a variety of    多种
be used to    习惯于

## Technical Terms

assembly language    汇编语言,装配语言
crystal [ˈkristl]    晶振
oscillator amplifier    振荡放大器
special function register    特殊功能寄存器
status register    状态寄存器
purpose register    目的寄存器
data pointer    数据指针
mov operations    mov 指令操作

## Lesson 28  ARM

The ARM architecture is a 32-bit advanced RISC design. The use of a simple design and a highly orthogonal, yet powerful instruction set allows it to attain good processing speeds at a low power consumption. As the demand for greater processing power in embedded devices keeps increasing, so does the popularity of ARM. The architecture is licensed by a large number a microprocessor manufacturers, leading to a wide choice and availability of ARM based CPUs and microcontrollers. Current ARM based CPUs are powerful enough to allow a large variety of DSP processing tasks to be performed without the use of a coprocessor.

The origins of ARM date back to the 1980's and the British computer manufacturer Acorn Computers. Acorn had enjoyed some success in the British market, getting large sales with it's "BBC Micro" computer. The BBC Micro was built by Acorn for the British Broadcasting Corporation's "BBC Computer Literacy Project", with the goal of educating the British population about what was considered the inevitable up-and-coming microcomputer revolution.

The BBC Micro was, for it's time, an advanced machine with good expandability and a rugged design—a necessary feature since it was widely used for teaching computer skills in classrooms. At it's hearth was a 2MHz MOS Technology 6502A. Over 1 million of these machines were sold for prices ranging from £300 to £400, allowing Acorn to expand it's business.

Meanwhile, the IBM PC had been launched, and began making headway in the business world. The Apple Lisa had also brought the first windowing system to the market. It was time for Acorn to move away from the 6502 CPU to a more capable alternative. There was however, one problem: Acorn's engineers had tested all 16-bit and 32-bit processors on the market, and none of them met their requirements. Notably, the Motorola 68000 and National Semiconductor 16032 processors were rejected because they had too big interrupt latencies, or were to slow and complex, respectively. Even at higher clock speeds, the very advanced NS 16032 was actually slower than the much simpler 6502.

Acorn decided to do the processor design themselves. They had sufficient cash reserves, meaning the project could take a while if needed, and some quite skilled research staff. What they didn't have was the experience nor the equipment or manpower to tackle a large design. At about the same time, the first formal research papers on RISC architectures started appearing from UC Berkeley and Stanford. This served as an encouragement: if a group of graduate students could come up with competitive new CPU designs, then Acorn surely had a good chance of succeeding.

A team consisting of Steve Furber, Roger Wilson and Robert Heaton set to work on the project, basing themselves on their large experience with writing system software for the 6502 and-from looking at the competing designs—the knowledge of how it should not be done. Work started in 1983 and finished in 1985, a mere 18 months and 5 man-years later. The Acorn RISC Machine 1 (ARM1) consisted of only 25,000 transistors. Much of the required tools were custom written in BASIC, to be run on the BBC Micro. In return, the first ARM prototypes were used as coprocessors for the Micro's, so the design tools for the complete system could finish faster.

The first experiences with writing system software for the ARM pointed to some inefficiencies, and those were resolved quickly in the ARM2. A multiply and multiply-accumulate instruction were added, both instructions which were usually the first to be thrown out in RISC designs. Putting them back in allowed the Acorn to do some real-time signal processing and have minimal sound capabilities, which was estimated to be important for a home and educational computer. While the design of the CPU for the next-generation Acorn computer was a success, the rest of the system was more problematic. Both the surrounding peripherals as the operating system took large amounts of time to finish, delaying the system by as much as 2 years. Meanwhile, the IBM PC had managed to set itself more or less as the standard, and the ARM's appearance with a new processor, and new operating system, and no software, was not met with much enthusiasm.

Acorn continued both the development of the ARM and the systems built on it, gradually allowing more software to appear, and boosting the performance of the ARM processors. The ARM3 added on-chip cache, which boosted the numbers of transistors to 300 000, still a tiny amount compared to other designs.

Some success was made selling the ARM CPU to other companies, for example as an embedded processor. One of the companies to buy the ARM was Apple, which built a new kind of device around it: the Apple Newton Personal Digital Assistant (PDA). The new device was not very successful at the time.

The collaboration with Apple, as well as the financial situation of Acorn, lead to the splitting off of the Advanced Research and Development section of Acorn into the newly founded Advanced RISC Machines Ltd. ARM Ltd. focused mainly on the design of the processor IP[1] cores. The actual silicon production is handled by the licensees.

While the sales and licensees of ARM Ltd. kept going up, Acorn's value kept going down until, eventually, it's share in ARM Ltd. was worth more than the company itself. It underwent a series of restructurings and repositioned itself as an IP developer in the DSP world, with the new name of Element 14.

Since ARM refers to both the architecture design as well as the processor IP cores implementing it, the naming can be confusion. The architectures are referred to as ARMvX with X being the version, and the cores just as ARMX, with X being the version.

### ARMv1, ARMv2 and ARMv2a

Introduced in 1985, the ARM1 was the first incarnation of the ARMv1 architecture, a 32-bit architecture, using 26-bit addressing. The ARMv2 and ARM2 added multiplication and multiply-accumulate capability, as well as coprocessor support.

The ARM3 (using ARMv2a) added a 4KB on-chip code and data cache.

### ARMv3, ARM6 and ARM7

ARMv3 moved the ARM architecture toward full 32-bit addressing, and also allowed for chips to optionally contain a Memory Management Unit (MMU), allowing the chip to run full operating systems. ARMv3M added optional long multiplication support ($32 \times 32$ bit = 64-bit), marked by a M in the processor type name.

### ARMv4, ARM7TDMI-S and Thumb[2]

The ARMv4 is a further evolution of the ARMv3 design, adding support for 16-bit load and stores, and dropped support for 26-bit addressing. It was introduced in 1995. A major step forwards in the ARM architecture was made by the introduction of Thumb in ARMv4T. Thumb is a new instruction set for the ARM architecture, using 16-bit instructions. It represents a significant enhancement to the ARM architecture, because it allows many operations to be performed at a significantly reduced code size. It is very useful in situations where bus or memory bandwidth or space is limited. The Thumb instructions are slightly less capable than the ARM counterparts, notably lacking conditional execution, shift/rotation support, and not allowing access to the entire register file. Because of the loss of orthogonality and cleanness, Thumb code is generally not written by hand.

Although Thumb reduces the size of each instruction by half, performing the same operations in Thumb usually requires some more instructions as in ARM mode. On average, one can expect a savings of 30%. It is possible to mix ARM routines with Thumb routines, allowing the performance critical parts of the code to be written in ARM code, and the rest of the system in Thumb, maximizing performance and minimizing code size.

The ARMv4T architecture is very widespread, because the ARM7TDMI core, which implements it, was widely licensed and adopted both in various microcontrollers as well as portable devices (such as the iPod or the Game Boy Advance). The ARM7TDMI includes wide (M)ultiply and (T)humb decoding, as well as the (D)ebug JTAG [3] and embedded (I)n-circuit emulator features.

The ARM7TDMI core is also available in a (S)ynthesizible version, allowing it to be compiled into other designs.

### ARMv5, ARM9EJS

The ARMv5 makes the optional Thumb and wide multiply features of the ARMv3 the default, and adds the following:

- Some extra instructions for smoother switching between ARM and Thumb states.
- Enhanced DSP support, including faster 16-bit multiplies and saturating arithmetic, and a faster MAC. Marked by an (E) in the chip name.
- Jazelle, allowing the ARM core to execute some Java bytecode instructions in hardware. Jazelle is an essential component in the Java support included in many mobile phones. Marked by a (J) in the chip name.

The ARM9 extends the 3-stage pipeline of the ARM7 to a 5-stage pipeline, allowing higher clock speeds. While most ARM7 cores peak around 80 MHz operation, ARM9 allows clock speeds in excess of 220MHz. The design also switched from a Von Neumann (shared data and instruction buses) to a Harvard architecture (separate data and instruction buses). It was released in 1997.

The ARM9EJ-S, released in 2000, was the first part to include Jazelle and is currently immensely popular in mobile phones and PDA's.

### ARMv5, ARM10

The ARM10 is an evolution of the ARMv5 architecture aimed at maximal performance. It extends the pipeline to 6 stages, and includes support for a floating point coprocessor (VFP). It is projected to run at speeds of around 260 MHz.

### ARMv6, ARM10, ARM11

The ARMv6 adds memory management features, a new multimedia instruction set, and a longer pipeline (8 stages), allowing even higher clock speeds, up to 335 MHz. The new multimedia instructions are SIMD3 and resemble the MMX, SSE and SSE2 instructions sets as used on common PCs. ARMv6 was introduced in 2001.

Neither the ARM10 nor ARM11 have had significant uptake in the market yet, and are not nearly as widely available as the ARM7TDMI and the ARM9EJS. Of course, this may change as time passes.

## New Words

expandability [ikˌspændə'biliti] n. 伸延性，扩展性，膨胀性
encouragement [in'kʌridʒmənt] n. 鼓励，奖励
inefficiency [ˌini'fiʃənsi] n. 无效率，无能
license ['laisəns] n. 许可(证)，执照；vt. 许可，特许
staff [stɑ:f] n. 棒，杖，杆，支柱，全体职员

## Phrases & Expressions

go up　　上升，增长，被兴建起来
put back　　放回原处，向后移，推迟，倒退，使后退
make headway　　取得进展
up-and-coming　　积极进取的，很有前途的，日见重要的

## Technical Terms

RISC　(Reduced Instruction Set Computer)[计] 精简指令集计算机
BBC　　英国广播公司
coprocessor　　n. [计] 协处理器
date back to　　从……时就有，回溯到，远在……(年代)
interrupt latency　　中断延迟
MMU　　[计] 存储器管理单元

## Notes

1. IP：Intellectual Property（知识产权），IP核将一些在数字电路中常用，但比较复杂的功能块，如 FIR 滤波器、SDRM 控制器、PCI 接口等设计做成一个"黑盒"或者可修改参数的模块，供设计者使用。IP核包括硬 IP 与软 IP。调用 IP 核能避免重复劳动，大大减少了设计人员的工作量。

2. Thumb：Thumb 指令集可以看作 ARM 指令压缩形式的子集，它是为减少代码量而提出的，具有 16bit 的代码密度。Thumb 指令体系并不完整，只支持通用功能，必要时仍需要使用 ARM 指令，如进入异常时。其指令的格式与使用方式与 ARM 指令集类似，而且使用并不频繁。

3. JTAG：JTAG 是联合测试工作组的简称，是名为标准测试访问端口和边界扫描结构的 IEEE 的标准 1149.1 的常用名称。此标准用于测试访问端口，使用边界扫描的方法来测试印制电路板。

## Exercises

1. Keywords

In the article, there are some important words which are the soul of this paper. After reading this paper, we can find some words to stand for this article. Now please find out key words of this paper.

2. Summary

After reading this paper, please write a summarize about ARM.

# 科技英语知识9：名词的翻译

几乎所有语言都存在一词多义的现象。在英汉词典中我们往往会查到一个英语单词的多重含义，因此在翻译科技文章时，必须结合语法知识和上下文的逻辑关系，尤其是结合所涉及的专业知识，才能对一个词的具体词义做出准确的判断。名词也不例外，在不同的专业和上下文中，往往具有不同的汉语翻译方式。例如，cell 在生物学中作"细胞"讲，在化工领域可作"电解槽"讲，在电学中是"电池"的意思；base 在日常作"基础"讲，在机械中可作"底座"讲，在化学中是"碱"的意思，三极管的 base 是"基极"，而三角形的 base 是"底边"。

在一篇文章中，既会因为与其他学科的交叉，同时也会由于上下文而涉及以上问题。例如，order 作名词使用时，就会在不同情况下有不同的意思：

- operational order(运算指令)；
- order of a differential equation(微分方程的阶)；
- order of matrix (矩阵的阶)；
- technical order(技术说明，技术规程)；
- be in/out of order(正常/发生故障)；
- in order to(为了)；
- order code(指令码)；
- order of connection(接通(连接)次序)；
- order of poles(极点的相重数)；
- working order(工序，加工单)；
- give an order for sth. (订货)；
- order of magnitude(数量级)。

可以看出，要想翻译正确，绝不能仅凭借日常生活用语中相对狭窄的知识面和词汇量，而应结合专业知识，在字典的帮助下，熟悉和积累常用英文专业词汇的汉语词义，切莫随意妄加推断。但字典不可能对每个词所有搭配的含义和翻译都一一列出，因而在必要时还需要对原文中的词汇做词义引申的意译处理，以避免译文生硬晦涩、意思含糊，甚至误解。

例：The study of neural network is one of the last frontiers of artificial intelligence.
对神经网络的研究是人工智能的最新领域之一。
last frontiers 由"最后的边疆"引申转译为"最新领域"。

# Unit 11　Computer Program Design

## Lesson 29　Introduction to C

C is a general-purpose programming language. It has been closely associated with the UNIX [1] operating system where it was developed, since both the system and most of the programs that run on it are written in C. The language, however, is not tied to any one operating system or machine; and although it has been called a "system programming language" because it is useful for writing compilers and operating systems, it has been used equally well to write major programs in many different domains.

Many of the important ideas of C stem from the language BCPL [2], developed by Martin Richards. The influence of BCPL on C proceeded indirectly through the language B, which was written by Ken Thompson in 1970 for the first UNIX system on the DEC PDP-7.

BCPL and B are "typeless" languages. By contrast, C provides a variety of data types. The fundamental types are characters, and integers and floating point numbers of several sizes. In addition, there is a hierarchy of derived data types created with pointers, arrays, structures and unions. Expressions are formed from operators and operands; any expression, including an assignment or a function call, can be a statement. Pointers provide for machine-independent address arithmetic.

C provides the fundamental control-flow constructions required for well-structured programs: statement grouping, decision making (if-else), selecting one of a set of possible values (switch), looping with the termination test at the top (while, for) or at the bottom (do), and early loop exit (break) [3].

Functions may return values of basic types, structures, unions, or pointers. Any function may be called recursively. Local variables are typically "automatic", or created anew with each invocation. Function definitions may not be nested but variables may be declared in a block-structured fashion. The functions of a C program may exist in separate source files that are compiled separately. Variables may be internal to a function, external but known only within a single source file, or visible to the entire program.

A preprocessing step performs macro substitution on program text, inclusion of other source files, and conditional compilation.

C is a relatively "low-level" language. This characterization is not pejorative; it simply means that C deals with the same sort of objects that most computers do, namely characters, numbers, and addresses. These may be combined and moved about with the arithmetic and logical operators implemented by real machines.

C provides no operations to deal directly with composite objects such as character strings, sets, lists or arrays. There are no operations that manipulate an entire array or string, although structures may be copied as a unit. The language does not define any storage allocation facility other than static definition and the stack discipline provided by the local variables of functions; there is no heap or garbage collection. Finally, C itself provides no input/output facilities; there are no READ or WRITE statements, and no built-in file access methods. All of these higher-level mechanisms must be provided by explicitly called functions. Most C implementations have included a reasonably standard collection of such functions.

Similarly, C offers only straightforward, single-thread control flow: tests, loops, grouping, and subprograms, but not multiprogramming, parallel operations, synchronization, or coroutines [4]. Although the absence of some of these features may seem like a grave deficiency(you mean I have to call a function to compare two character strings?), keeping the language down to modest size has real benefits. Since C is relatively small, it can be described in small space, and learned quickly. A programmer can reasonably expect to know and understand and indeed regularly use the entire language.

In 1983, the American National Standards Institute (ANSI) established a committee to provide a modern, comprehensive definition of C. The resulting definition, the ANSI standard, or " ANSI C", was completed in late 1988. Most of the features of the standard are already supported by modern compilers.

The standard is based on the original reference manual. The language is relatively little changed; one of the goals of the standard was to make sure that most existing programs would remain valid, or, failing that, that compilers could produce warnings of new behavior. For most programmers, the most important change is the new syntax for declaring and defining functions. A function declaration can now include a description of the arguments of the function; the definition syntax changes to match. This extra information makes it much easier for compilers to detect errors caused by mismatched arguments; in our experience, it is a very useful addition to the language.

There are other small-scale language changes. Structure assignment and enumerations, which had been widely available, are now officially part of the language. Floating-point computations may now be done in single precision. The properties of arithmetic, especially for unsigned types, are clarified. The preprocessor is more elaborate. Most of these changes will have only minor effects on most programmers.

A second significant contribution of the standard is the definition of a library to accompany C. It specifies functions for accessing the operating system (for instance, to read and write files), formatted input and output, memory allocation, string manipulation, and the like. A collection of standard headers provides uniform access to declarations of functions in data types. Programs that use this library to interact with a host system are assured of compatible behavior. Most of the library is closely modeled on the "tandard I/O library" of the UNIX system.

Because the data types and control structures provided by C are supported directly by most computers, the run-time library required to implement self-contained programs is tiny. The standard library functions are only called explicitly, so they can be avoided if they are not needed. Most can be written in C, and except for the operating system details they conceal, are themselves portable.

Although C matches the capabilities of many computers, it is independent of any particular machine architecture. With a little care it is easy to write portable programs, that is, programs that can be run without change on a variety of hardware. The standard makes portability issues explicit, and prescribes a set of constants that characterize the machine on which the program is run.

C is not a strongly-typed language, but as it has evolved, its type-checking has been strengthened. The original definition of C frowned on, but permitted, the interchange of pointers and integers; this has long since been eliminated, and the standard now requires the proper declarations and explicit conversions that had already been enforced by good compilers. The new function declarations are another step in this direction. Compilers will warn of most type errors, and there is no automatic conversion of incompatible data types. Nevertheless, C retains the basic philosophy that programmers know what they are doing; it only requires that they state their intentions explicitly.

C, like any other language, has its blemishes. Some of the operators have the wrong precedence; some parts of the syntax could be better. Nonetheless, C has proven to be an extremely effective and expressive language for a wide variety of programming applications.

## New Words

hierarchy ['haiərɑːki] *n.* [计]数据层次,体系;[制]系统,谱系
termination [ˌtəːmiˈneiʃən] *n.* 结局,结束;终止,末端;终点
pointer ['pɔintə] *n.* 指示者;指示;(钟、表、仪表、天平、秤等的)指针
substitution [ˌsʌbstiˈtjuːʃən] *n.* 代(用),代(更)替,置(替,变)换;[化]取代
pejorative ['piːdʒərətiv] *adj.* 贬低的,恶化的,变坏的;使带有轻蔑
relatively ['relətivli] *adv.* 相对地;比较地
enumeration [iˌnjuːməˈreiʃən] *n.* 计算,列举,调查,细目,详表
elaborate [iˈlæbərət] *adj.* 精致的,精巧的
invocation [ˌinvəuˈkeiʃən] *n.* 援引;发动
modest ['mɔdist] *adj.* 谨慎的,谦虚的;谦让的;适度的,适中的

## Technical Terms

coroutine [ˌkəruːˈtiːn] *n.* 协同程序
evolve [iˈvɔlv] *vt.* 使发展,(使)进化,演化,伸出,推理,引出

recursively [ri'kəsivli] *adj.* 回归的，递归的
syntax ['sintæks] *n.* 句法；句子结构学；措辞法；字句排列法
preprocess [pri:'prəuses] *vt.* 预加工；预处理
system programming language  系统编程语言
machine-independent  独立于机器的，与机器无关的
control-flow  控制流
floating-point  浮点

## Notes

1. UNIX：一个强大的多用户、多任务操作系统，支持多种处理器架构，按照操作系统的分类，属于分时操作系统，最早由 Ken Thompson、Dennis Ritchie 和 Douglas McIlroy 于 1969 年在 AT&T 公司的贝尔实验室开发。

2. BCPL：Basic (Bootstrap) Combined Programming Language，基本（自展）组合程序设计语言。

3. C provides the fundamental control-flow constructions required for well-structured programs: statement grouping, decision making (if-else), selecting one of a set of possible values (switch), looping with the termination test at the top (while, for) or at the bottom (do), and early loop exit (break).

C 为实现结构良好的程序提供了基本的控制流结构：语句组、判断（if-else）、选择一个可能的路径（switch）、终止测试在顶端进行（while、for）、底端进行（do）的循环和提前跳出循环（break）。

4. Similarly, C offers only straightforward, single-thread control flow: tests, loops, grouping, and subprograms, but not multiprogramming, parallel operations, synchronization, or coroutines.

类似地，C 只提供单线程的控制流：测试、循环、组合和子程序，不提供多道程序设计、并行操作、同步和协同例程。

# Lesson 30  HDL Language

A digital system can be described at different levels of abstraction and from different points of view. An HDL should faithfully and accurately model and describe a circuit, whether already built or under development, from either the structural or behavioral views, at the desired level of abstraction. Because HDLs are modeled after hardware, their semantics and use are very different from those of traditional programming languages. The following subsections discuss the need, use and design of an HDL.

### Limitations of traditional programming languages

There are wide varieties of computer programming languages, from Fortran to C to Java.

Unfortunately, they are not adequate to model digital hardware. To understand their limitations, it is beneficial to examine the development of a language. A programming language is characterized by its syntax and semantics. The syntax comprises the grammatical rules used to write a program, and the semantics is the "meaning associated" with language constructs. When a new computer language is developed, the designers first study the characteristics of the underlying processes and then develop syntactic constructs and their associated semantics to model and express these characteristics. Associated semantics to model and express these characteristics.

Most traditional general-purpose programming languages, such as C, are modeled after a sequential process. In this process, operations are performed in sequential order, one operation at a time. Since an operation frequently depends on the result of an earlier operation, the order of execution cannot be altered at will. The sequential process model has two major benefits. At the abstract level, it helps the human thinking process to develop an algorithm step by step. At the implementation level, the sequential process resembles the operation of a basic computer model and thus allows efficient translation from an algorithm to machine instructions.

The characteristics of digital hardware, on the other hand, are very different from those of the sequential model. A typical digital system is normally built by smaller parts, with customized wiring that connects the input and output ports of these parts. When a signal changed, the parts connected to the signal are activated and a set of new operations is initiated accordingly. These operations are performed concurrently, and each operation will take a specific amount of time, which represents the propagation delay of a particular part, to complete. After completion, each part updates the value of the corresponding output port. If the value is changed, the output signal will in turn activate all the connected parts and initiate another round of operations. This description shows several unique characteristics of digital systems, including the connections of parts, concurrent operations, and the concept of propagation delay and timing. The sequential model used in traditional programming languages cannot capture the characteristics of digital hardware, and there is a need for special languages (HDLs) that are designed to model digital hardware.

### Use of an HDL program

To better understand HDL, it is helpful to examine the use of an HDL program. In a traditional programming language, a program is normally coded to solve a specific problem. It takes certain input values and generates the output accordingly. The program is first compiled to machine instructions and then run on a host computer. On the other hand, the application of an HDL program is very different. The program plays three major roles:

Formal documentation. A digital system normally starts with a word description. Unfortunately, since human language is not precise, the description is frequently incomplete and ambiguous, and the same description may be subject to different interpretations. Because the semantics and syntax of an HDL are defined rigorously, a system specified in an HDL program is explicit and precise. Thus, an HDL program can be used as a formal system specification and documentation among various designers and users.

Input to a simulator. Simulation is used to study and verify the operation of a circuit without constructing the system physically. An HDL simulator provides a framework to model the concurrent operations in a sequential host computer, and has specific knowledge of the language's syntactic constructs and the associated semantics [1]. An HDL program, combined with test vector generation and a data collection code, forms a testbench, which becomes the input to the HDL simulator. During execution, the simulator interprets HDL code and generates responses accordingly.

Input to a synthesizer. The modern development flow is based on the refinement process, which gradually converts a high-level behavioral description to a low-level structural description. Some refinement steps can be performed by synthesis software. The synthesis software takes an HDL program as its input and realizes the circuit from the components of a given library. The output of the synthesizer is a new HDL program that represents the structural description of the synthesized circuit.

### Design of a modern HDL

The fundamental characteristics of a digital circuit are defined by the concepts of entity, connectivity, concurrency and timing. Entity is the basic building block, modeling after a part of a real circuit. It is self-contained and independent, and has no implicit information about other entities. Connectivity models the connecting wires among the parts. It is the way that entities interact with one another. Since the connections of a system are seldom formed as a single thread, many entities may be active at the same time and many operations are performed in parallel. Concurrency describes this type of behavior. Timing is related to concurrency. It specifies the initiation and completion of each operation and implicitly provides a schedule and order of multiple operations.

The goal of an HDL is to describe and model digital systems faithfully and accurately. To achieve this, the cornerstone of the language should be based on the model of hardware operation, and its semantics should be able to capture the fundamental characteristics of the circuits. A digital system can be described at four different levels of abstraction and from three different points of view. Although these descriptions have similar fundamental characteristics, their detailed representations and models vary significantly. Ideally, we wish to develop a single HDL to cover all the levels and all the views. However, this is hardly feasible because the vast differences between abstraction levels and views will make the language excessively complex. Modern HDLs normally cover descriptions in structural and behavior views, but not in physical view. They provide constructs to support modeling at the gate and RT levels, and to a limited degree, at processor and transistor levels. The highlights of modern HDLs are as follows:

The language semantics encapsulate the concepts of entity, connectivity, concurrency, and timing.

The language can effectively incorporate propagation delay and timing information.

The language consists of constructs that can explicitly express the structural implementation (a block diagram) of a circuit.

The language incorporates constructs that can describe the behavior of a circuit, including constructs that resemble the sequential process of traditional languages, to facilitate abstract behavioral description.

The language can efficiently describe the operations and structures at the gate and RT levels.

The language consists of constructs to support a hierarchical design process.

## New Words

abstraction [æb'strækʃən] n. 抽象概念,提取,萃取,抽出
semantics [si'mæntiks] n. [语]语义学;[哲]语义哲学;语义学派
grammatical [grə'mætikəl] adj. 语法的,语法上的
syntactic [sin'kæktik] adj. 合成的;句法的
resemble [ri'zembl] vt. 像;类似
incomplete [ˌinkəm'pliːt] adj. 不完全的,未完成的,不完备的;不足的
ambiguous [æm'bigjuəs] adj. 有两种或多种意思的,含糊的,模棱两可的
rigorously ['rigərəsli] adv. 严厉地,残酷地
synthesizer ['sinθisaizə] n. 合成者,合成物;[电]合成器,综合器
refinement [ri'fainmənt] n. 精(提)炼,提纯,精制;净化,纯净,精美
self-contained ['selfkən'teind] adj. 独立的;自足的,自恃的;自治的;自制的
cornerstone ['kɔːnəstəun] n. 隅石,墙角石,奠基石;基石

## Phrases & Expressions

take...as  把……作为

## Technical Terms

testbench  测试平台
data collection  数据采集
entity  实体
connectivity  连接体
concurrency  并发体
gate and RT levels  门级和寄存器传输级
machine instructions  机器指令
propagation delay  传输延迟

**Note**

1. An HDL simulator provides a framework to model the concurrent operations in a sequential host computer, and has specific knowledge of the language's syntactic constructs and the associated semantics.

HDL仿真器提供了一个在时序宿主计算机上建立并行操作模型的架构,它有语言句法结构和相关语义的特定知识库。

## Lesson 31  Perl

Perl is a general-purpose programming language originally developed for text manipulation and now used for a wide range of tasks including system administration, web development, network programming, games, and GUI development.

The language is intended to be practical (easy to use, efficient, complete) rather than beautiful (tiny, elegant, minimal). Its major features include support for multiple programming paradigms (procedural, object-oriented, and functional styles), reference counting memory management (without a cycle-detecting garbage collector), built-in support for text processing, and a large collection of third-party modules.

According to Larry Wall, Perl has two slogans. The first is "There's more than one way to do it", commonly known as TMTOWTDI. The second slogan is "Easy things should be easy and hard things should be possible [1]."

### Features

The overall structure of Perl derives broadly from C. Perl is procedural in nature, with variables, expressions, assignment statements, brace-delimited code blocks, control structures, and subroutines.

Perl also takes features from shell programming. All variables are marked with leading sigils, which unambiguously identify the data type (for example, scalar, array, hash) of the variable in context[2]. Importantly, sigils allow variables to be interpolated directly into strings. Perl has many built-in functions that provide tools often used in shell programming (although many of these tools are implemented by programs external to the shell) such as sorting, and calling on system facilities.

Perl takes lists from Lisp, associative arrays (hashes) from AWK[3], and regular expressions from sed. These simplify and facilitate many parsing, text-handling, and data-management tasks.

In Perl 5, features were added that support complex data structures, first-class functions (that is, closures as values), and an object-oriented programming model. These include references, packages, class-based method dispatch, and lexically scoped variables, along with compiler directives (for example, the strict program). A major additional feature introduced with Perl 5 was the ability to package code as reusable modules. Larry Wall later stated that "The whole intent of Perl 5's module system was to encourage the growth of Perl culture rather than the Perl core."

All versions of Perl do automatic data typing and memory management. The interpreter knows the type and storage requirements of every data object in the program; it allocates and frees storage for them as necessary using reference counting (so it cannot deal locate circular data structures without manual intervention). Legal type conversions—for example, conversions from number to string—are done automatically at run time; illegal type conversions are fatal errors.

### Design

The design of Perl can be understood as a response to three broad trends in the computer industry: falling hardware costs, rising labor costs, and improvements in compiler technology. Many earlier computer languages, such as FORTRAN and C, were designed to make efficient use of expensive computer hardware. In contrast, Perl is designed to make efficient use of expensive computer programmers.

Perl has many features that ease the programmer's task at the expense of greater CPU and memory requirements. These include automatic memory management; dynamic typing; strings, lists, and hashes; regular expressions; introspection; and an eval () function.

Wall was trained as a linguist, and the design of Perl is very much informed by linguistic principles. Examples include Huffman coding (common constructions should be short), good end-weighting (the important information should come first), and a large collection of language primitives. Perl favors language constructs that are concise and natural for humans to read and write, even where they complicate the Perl interpreter.

Perl syntax reflects the idea that "things that are different should look different". For example, scalars, arrays, and hashes have different leading sigils. Array indices and hash keys use different kinds of braces. Strings and regular expressions have different standard delimiters. This approach can be contrasted with languages such as Lisp, where the same S-expression construct and basic syntax are used for many different purposes.

Perl does not enforce any particular programming paradigm (procedural, object-oriented, functional, and others) or even require the programmer to choose among them.

There is a broad practical bent to both the Perl language and the community and culture that surround it. Perl is a language for getting your job done. One consequence of this is that Perl is not a tidy language. It includes many features, tolerates exceptions to its rules, and employs heuristics to resolve syntactical ambiguities. Because of the forgiving nature of the compiler, bugs can sometimes be hard to find.

In addition to Larry Wall's two slogans mentioned above, Perl has several mottos that convey aspects of its design and use, including "Perl: the Swiss Army Chainsaw of

Programming Languages" and "No Unnecessary Limits". Perl has also been called "The Duct Tape of the Internet".

No written specification or standard for the Perl language exists, and there are no plans to create one for the current version of Perl. There has been only one implementation of the interpreter, and the language has evolved along with it. That interpreter, together with its functional tests, stands as a de facto specification of the language.

### Applications

Perl has many and varied applications, compounded by the availability of many standard and third-party modules.

Perl has been used since the early days of the Web to write CGI[4] scripts. It is known as one of "the three Ps" (along with Python and PHP), the most popular dynamic languages for writing Web applications. It is also an integral component of the popular LAMP solution stack for web development. Large projects written in Perl include Slash, Bugzilla, RT, TWiki, and Movable Type. Many high-traffic websites use Perl extensively. Examples include bbc.co.uk, Amazon.com, LiveJournal, Ticketmaster, Slashdot, Craigslist, Zappos.com and IMDb.

Perl is often used as a glue language, tying together systems and interfaces that were not specifically designed to interoperate, and for "data munging", that is, converting or processing large amounts of data for tasks such as creating reports. In fact, these strengths are intimately linked. The combination makes Perl a popular all-purpose language for system administrators, particularly because short programs can be entered and run on a single command line.

With a degree of care, Perl code can be made portable across Windows and UNIX. Portable Perl code is often used by suppliers of software (both COTS and bespoke) to simplify packaging and maintenance of software build and deployment scripts.

Graphical user interfaces (GUIs) may be developed using Perl. For example, Perl/Tk is commonly used to enable user interaction with Perl scripts. Such interaction may be synchronous or asynchronous using callbacks to update the GUI. For more information about the technologies involved, see Tk, Tcl, and WxPerl.

Perl is also widely used in finance and bioinformatics, where it is valued for rapid application development and deployment and for its capability to handle large data sets.

## New Words

manipulation [məˌnipjuˈleiʃən] n. 操作；控制；处理；计算，运算
paradigm [ˈpærədaim] n. 典范；范例；示例
procedural [prəˈsiːdʒərəl] adj. 过程式；[律]程序上的，程序性的
unambiguously [ˈʌnæmˈbigjuəsli] adv. 明白地，不含糊地
interpolate [inˈtəːpəuleit] vt., vi. 篡改，插入；[数]插值，内插，内推
parsing [ˈpɑːziŋ] n. 分(剖)析，分解

closure ['kləuʒə] *n.* 关闭,停业;截止;末尾
package ['pækidʒ] *n.* 封装,包裹,包;捆,束,件头
introspection [,intrəu'spekʃən] *n.* 内省,反省
delimiter [di:'limitə] *n.* 分界符;定义符;限定器
motto ['mɔtəu] *n.* (*pl.* mottos, mottoes)标语;座右铭,箴言
portable ['pɔ:təbl] *adj.* 便于携带的;手提式的,轻便的,可移动的
bioinformatics  生物信息学

## Phrases & Expressions

mark with    以……表明,以……为标记
tie together    使……捆在一起
contrasted with    与……截然不同
at the expense of    在损失某事物的情况下
call on    调用

## Technical Terms

subroutine [,sʌbru:'ti:n] *n.* [电子]子程序
sort [sɔ:t] *n.* 排序,种类,类别,品种
dispatch [dis'pætʃ] *vt.* 调用,调度,派遣;派出;(火速)发送(信件,公文等)
intervention [,intə(:)'venʃən] *n.* 插入,介入,调节,调停,干涉,妨碍
deallocate [di:ælə'keiʃən] *vt.* [计]解除分配,再分配,解除分配,释放
callback    回调函数
lexically scoped variables    词法范围变量
object-oriented    面向对象
functional styles    函数式
assignment statement    赋值语句
leading sigils    前缀,前置标记
built-in functions    内置功能,内部函数
compiler directive    编译指示
GUI(graphical user interface)    图形用户界面
written specification    书面规范
type conversion    类型变换
regular expression    正则表达式
end-weighting    句尾重心问题
labor costs    人工成本
hash key    散列键
array indices    数组索引
class-based    基类

## Notes

1. Easy things should be easy and hard things should be possible.
让简单的事情变得更容易,让困难的事情成为可能。

2. All variables are marked with leading sigils, which unambiguously identify the data type (for example, scalar, array, hash) of the variable in context.
所有变量都标记上前置标识符,这样可以精确地定义上下文中变量的数据类型(如标量、数组、散列)。

3. AWK:一种编程语言,用于在 Linux/UNIX 系统中对文本和数据进行处理。数据可以来自标准输入、一个或多个文件,或其他命令的输出。它支持用户自定义函数和动态正则表达式等先进功能,是 Linux/UNIX 系统中的一个强大编程工具。它在命令行中使用,但更多地是作为脚本来使用。AWK 的处理文本和数据的方式是这样的,它逐行扫描文件,从第一行到最后一行,寻找匹配的特定模式的行,并在这些行上进行你想要的操作。如果没有指定处理动作,则把匹配的行显示到标准输出(屏幕),如果没有指定模式,则所有被操作所指定的行都被处理。AWK 分别代表其作者姓氏的第一个字母,这三个人分别是 Alfred Aho、Peter Weinberger、Brian Kernighan。

4. CGI:Common Gateway Interface,公共网关接口,HTTP 服务器与用户或其他机器上的程序进行"交谈"的一种工具,其程序必须运行在网络服务器上。

## Exercises

### 1. Keywords

In the article, there are some important words which are the soul of this paper. After reading this paper, we can find some words to stand for this article. Now please find out key words of this paper.

### 2. Summary

After reading this paper, please write a summary about Perl.

# 科技英语知识 10：论文的标题和摘要

科技论文的标题和摘要是概括全文和吸引读者的重要部分。为了便于国际学术交流和文献索引，即使是在国内发表的论文往往也需要加上英文题目和摘要。

科技论文的标题应该能隐含文章的主题与大意，以便于索引。一般情况下，标题力求简明扼要，没有主、谓、宾的固定结构，长度在 10 个词以内，最多不超过 15 个词，句式应避免问句格式，且不要使用不标准的缩略语。作者和作者的单位放在标题的下方。

摘要是全文的缩影和总结。通过阅读摘要，读者能很快了解全文的梗概，决定是否阅读全文。摘要的长度一般在 100 个单词左右或稍长一些，要保证重点突出、言简意赅、内容完整、结构严谨，避免过分简单和使用不标准的缩略语。

此外，摘要中不应使用问句或感叹句，常用的句型和例句有：

This paper treats(introduces) an important problem in …

A new technique(method, manner) on … is shown in this paper.

This paper provides(shows, develops, extends, describes) a new approach for …

The author describes the technique of …

It is suggested that some basic steps be taken in order to …

This paper presents a thorough study of the input/output stability of relay pulse sender.

The purpose of this article is to explore the relation between sampling period and stability.

关键词是论文中最重要且出现频率最高的词或词组。列出关键词有助于读者对全文的理解，同时便于查阅和检索。关键词一般使用名词形式，词数为 3~6 个。

中文摘要和英文摘要的写作原则基本相同。同一篇论文的中英文摘要必须保证内容上的一致，分句方式和表达方式则可以适当灵活使用。

# Unit 12　Communication Network Technology

## Lesson 32　Basic Telecommunications Network

The basic purpose of a telecommunications network is to transmit user information in any form to another user of the network. These users of public networks, for example, a telephone network, are called subscribers. User information may take many forms, such as voice or data, and subscribers may use different access network technologies to access the network, for example, fixed or cellular telephones. We will see that the telecommunications network consists of many different networks providing different services, such as data, fixed, or cellular telephony service. In the following section we introduce the basic functions that are needed in all networks no matter what services they provide.

The three technologies needed for communication through the network are (1) transmission, (2) switching, and (3) signaling. Each of these technologies requires specialists for their engineering, operation, and maintenance.

### Transmission

Transmission is the process of transporting information between end points of a system or a network. Transmission systems use four basic media for information transfer from one point to another:

(1) Copper cables, such as those used in LANs and telephone subscriber lines;

(2) Optical fiber cables, such as high-data-rate transmission in telecommunications networks;

(3) Radio waves, such as cellular telephones and satellite transmission;

(4) Free-space optics, such as infrared remote controllers.

In a telecommunications network, the transmission systems interconnect exchanges and, taken together, these transmission systems are called the transmission or transport network. Note that the number of speech channels (which is one measure of transmission capacity) needed between exchanges is much smaller than the number of subscribers because only a small fraction of them have calls connected at the same time.

### Switching

In principle, all telephones could still be connected to each other by cables as they were in the very beginning of the history of telephony. However, as the number of telephones grew, operators soon noticed that it was necessary to switch signals from one wire to another. Then only a few cable connections were needed between exchanges because the number of simultaneously ongoing calls is much smaller than the number of telephones (Figure 32-1). The first switches were not automatic so switching was done manually using a switchboard.

Stronger developed the first automatic switch (exchange) in 1887. At that time, switching had to be controlled by the telephone user with the help of pulses generated by a dial. For many decades exchanges were a complex series of electromechanical selectors, but during the last few decades they have developed into software-controlled digital exchanges. Modern exchanges usually have quite a large capacity — tens of thousands subscribers — and thousands of them may have calls ongoing at the same time.

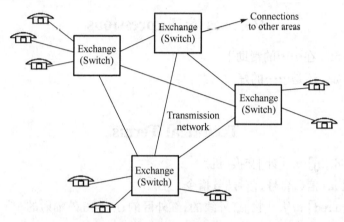

Figure 32-1    A basic telecommunications network

### Signaling

Signaling is the mechanism that allows network entities (customer premises or network switches) to establish, maintain, and terminate sessions in a network. Signaling is carried out with the help of specific signals or messages that indicate to the other end what is requested of it by this connection. Some examples of signaling examples on subscriber lines are as follows:

● Off-hook condition: The exchange notices that the subscriber has raised the telephone hook (dc loop is connected) and gives a dial tone to the subscriber.

● Dial: The subscriber dials digits and they are received by the exchange.

● On-hook condition: The exchange notices that the subscriber has finished the call (subscriber loop is disconnected), clears the connection, and stops billing.

Signaling is naturally needed between exchanges as well because most calls have to be connected via more than just one exchange. Many different signaling systems are used for the interconnection of different exchanges. Signaling is an extremely complex matter in a telecommunications network. Imagine, for example, a foreign GSM[1] subscriber switching his telephone on in Hong Kong. In approximately 10 seconds he is able to receive calls directed to him. Information transferred for this function is carried in hundreds of signaling messages between exchanges in international and national networks.

## New Words

subscriber [sʌbs'kraibə] n. 签署者,捐献者,用户;[经] 定户
telephony [ti'lefəni] n. 电话通信,电话制造;[电] 电话学
ongoing ['ɔngəuiŋ] adj. 前进的,进行的,不间断的
electromechanical [i,lektrəumi'kænikəl] adj. 电机的,电机学的;[计] 机电的

## Phrases & Expressions

with the help of　在……的帮助下
as well　倒不如,还是……的好

## Technical Terms

switching ['switʃiŋ] n. [计]交换(机)
signal ['signl] n. 信(暗)号;信号器;指令
infrared ['infrə'red] adj. [物]红外线的;红外区的;产生红外辐射的
cellular telephone　移动电话
optical fiber cable　光导纤维电缆
infrared remote controller　红外线远程控制器
off-hook　摘机
on-hook　挂机

## Note

1. GSM：Global System for Mobile Communications,[电信]全球移动通信系统,也称泛欧数字式移动通信系统,是一个根据欧洲电信标准协会颁布的 GSM 技术规范建造的国际无线蜂窝网。

# Lesson 33　Third Generation Cellular Systems

The main forces behind development of the third generation systems (3G) have been driven by the second generation systems' low performance data services, incompatible service in different parts of the world, and lack of capacity. In the 1990s, the ITU[1] started a project to develop a future global 3G system, which is known today as International Mobile Communications (IMT)-2000.

## IMT-2000

The IMT-2000 system was designed to be a global system for third generation mobile communications. It was developed by the ITU, which called it previously future public land mobile telecommunications system. Many problems have prevented the achievement of mutual understanding among countries regarding this system. Among the problems are frequency allocation in different continents, existing different second generation infrastructures, and different political interests. As a consequence a common understanding about detailed implementation technology was not achieved and IMT-2000 will not be a globally compatible technology; it will instead act as an umbrella for compatible services provided by different underlying technologies [2].

Even though third generation systems will use different technologies, but the development of mobile terminal technology will partly solve the incompatibility problem for users. With the same terminal we will be able use different networks and the services they provide. The most important network technology for 3G is UMTS.

## UMTS

UMTS is a European concept for integrated mobile services and it is based on the GSM and GPRS. Its goal is to provide a wide range of mobile services wherever the user is located. For UMTS cordless, cellular and satellite interfaces are defined. It will provide multimedia service with data rates up to 2Mbps for steady MSS [3] and up to 384Kbps for moving MSS.

The cellular radio access method for UMTS[4] approved ETSI[5] is wideband CDMA (WCDMA). The basic operating principle is the same as in CDMA[6]. The new frequency band at the 2GHz range is allocated for UMTS. The channel bandwidth is 5MHz, and each channel is used by all cells. The core network of UMTS is based on the core network of GSM and GPRS. The UMTS BSS [7] can be added to the GSM/GPRS network to operate in parallel with GSM base stations. Even handovers between UMTS and GSM/GPRS are supported.

## CDMA2000

The main 3G technology for the United States is based on second generation IS-95 CDMA. CDMA2000 is specified to use a sophisticated modulation scheme to increase the data rate over an ordinary 1.25MHz CDMA channel. The problem with 3G systems in the United States is that a much smaller frequency band is available for 3G service than in areas following European frequency division.

## W-CDMA

W-CDMA[8] was developed by NTT DoCoMo as the air interface for their 3G network FOMA (Frontier of Mobile multimedia Access). Later NTT DoCoMo submitted the specification to the International Telecommunication Union (ITU) as a candidate for the international 3G standard known as IMT-2000. The ITU eventually accepted W-CDMA as part of the IMT-2000 family of 3G standards, as an alternative to CDMA2000, EDGE[9] (Enhanced Data rates for GSM Evolution), and the short range DECT(Digital Enhanced

Cordless Telecommunications) system. Later, W-CDMA was selected as the air interface for UMTS, the 3G successor to GSM.

### TD-SCDMA

TD-SCDMA[10] is being pursued in the People's Republic of China by the Chinese Academy of Telecommunications Technology (CATT), Datang and Siemens AG, in an attempt not to be dependent on Western technology. This is likely primarily for practical reasons, other 3G formats require the payment of patent fees to a large number of western patent holder.

TD-SCDMA is based on spread spectrum technology which makes it unlikely that it will be able to escape completely the payment of license fees to western patent holders. The launch of a national TD-SCDMA network was initially projected by 2005 but has still not been achieved; the latest stage of "commercial trials" across eight cities was launched on April 1, 2008 and will eventually include 60 000 users. On January 7, 2009 China granted TD-SCDMA 3G license to China Mobile.

## New Words

force [fɔːs] *n.* 武力,兵力;实施;[常用复]部队,军队
capacity [kəˈpæsiti] *n.* 容量,容积;吸收力,收容力,生产力,通过率
mutual [ˈmjuːtjuəl, ˈmjuːtʃuəl] *adj.* 相互的;彼此的;[口]共有的,共同的
handover [ˈhændəuvə] *n.* 移交;转交;提交
continent [ˈkɔntinənt] *n.* 大陆;陆地;洲
infrastructure [ˈinfrəˈstrʌktʃə] *n.* 基础;基础结构
implementation [ˌimplimenˈteiʃən] *n.* 执行,履行;落实,供给工具

## Technical Terms

cellular systems　移动系统
mobile terminal　移动终端

## Notes

1. ITU: International Telecommunications Union,国际电信同盟。

2. As a consequence a common understanding about detailed implementation technology was not achieved and IMT-2000 will not be a globally compatible technology; it will instead act as an umbrella for compatible services provided by different underlying technologies.
由于具体实施的技术没有达成一致,IMT-2000将不会是全球兼容的技术,这将会成为不同的底层技术所提供的兼容服务的保护伞。

3. MSS: Maxitum Segment Size,最大报文长度,是TCP里的一个概念,就是TCP数据包

每次能够传输的最大数据分段。为了达到最佳的传输效能，TCP 在建立连接时通常要协商双方的 MSS 值，TCP 在实现的时候往往用 MTU 值代替这个值（需要减去 IP 数据包包头的大小 20Bytes 和 TCP 数据段的包头 20Bytes），所以 MSS 值往往为 1460。通信双方会将双方提供的 MSS 值的最小值确定为这次连接的最大 MSS 值。

4. UMTS：Universal Mobile Telecommunications System，通用移动通信系统。

5. ETSI：European Telecommunication Standards Institute，欧洲电信标准协会。

6. CDMA：Code Division Multiple Access，码分多址，是基于扩频技术的一种无线通信技术。

7. BSS：British Standard Specification(s)，英国标准（技术）规格。

8. WCDMA：Wideband Code Division Multiple Access，宽带码分多址移动通信系统。

9. EDGE：Enhanced Data rates for GSM Evolution，演进的 GSM 增强数据传输速率。

10. TD-SCDMA：Time Division-Synchronous Code Division Multiple Access，时分同步的码分多址技术。

## Lesson 34   Multimedia Network

Many applications, such as video mail, video conferencing, and collaborative work systems, require networked multimedia. In these applications, the multimedia objects are stored at a server and played back at the client's sites. Such applications might require broadcasting multimedia data to various remote locations or accessing large depositories of multimedia sources. Multimedia networks require a very high transfer rate or bandwidth, even when the data is compressed. Traditional networks are used to provide error-free transmission. However, most multimedia applications can tolerate errors in transmission due to corruption or packet loss without retransmission or correction [1]. In some cases, to meet real-time delivery requirements or to achieve synchronization, some packets are even discarded. As a result, we can apply lightweight transmission protocols to multimedia networks. These protocols cannot accept retransmission, since that might introduce unacceptable delays.

Multimedia networks must provide the low latency required for interactive operation. Since multimedia data must be synchronized when it arrives at the destination site, networks should provide synchronized transmission with low jitter.

In multimedia networks, most communications are multipoint as opposed to traditional point-to-point communication. For example, conferences involving more than two participants need to distribute information in different media to each participant. Conference networks use multicasting and bridging distribution methods. Multicasting replicates a single input signal and delivers it to multiple destinations. Bridging combines multiple input signals into one or more output signals, which then deliver to the participants.

Traditional networks do not suit multimedia Ethernet, which provides only 10 Mbps, its access time is not bounded, and its latency and jitter are unpredictable. Token-ring networks provide 16Mbps and are deterministic. From this point of view, they can handle multimedia. However, the predictable worst case access latency can be very high.

A fiber distributed data interface (FDDI) network provides 100 Mbps bandwidth, sufficient for multimedia. In the synchronized mode, FDDI has a low access latency and low jitter. It also guarantees a bounded access delay and a predictable average bandwidth for synchronous traffic. However, due to the high cost, FDDI networks are used primarily for backbone networks, rather than networks of workstations.

Less expensive alternatives include enhanced traditional networks. Fast Ethernet, for example, provides up to 100 Mbps bandwidth. Priority token ring is another system. In priority token ring networks, the multimedia traffic is separated from regular traffic by priority. Figure 34-1 shows a priority token ring. The bandwidth manager plays a crucial role by tracking sessions, determining ratio priority, and registering multimedia sessions. Priority token ring (PTR) works on existing networks and does not require configuration control. The admission control in PTR guarantees bandwidth to multimedia sessions. However, regular traffic experiences delays.

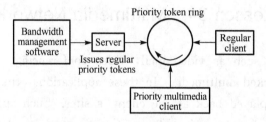

Figure 34-1  Priority taken ring

Example, Assume a priority token ring network at 16 Mbps that connects 32 nodes. When no priority scheme is set, each node gets an average of 0.5 Mbps of bandwidth. When half the bandwidth (8 Mbps) is dedicated to multimedia, the network can handle about 5MPEG sessions (at 1.5 Mbps). In that case, the remaining 27 nodes can expect about 8 Mbps divided by 27, about half of what they would get without priority enabled.

They are three priority ring schemes for their applicability to video conferencing applications:

(1) Equal priority for video and asynchronous packets.

(2) Permanent high priority for video packets and permanent low priority for asynchronous packets.

(3) Time-adjusted high priority for packets and permanent low priority for asynchronous packets.

The first scheme, which entails direct competition between videoconference and synchronous stations, achieves the lowest network delay for asynchronous frame. However, it reduces the videoconference quality. The second scheme, in which videoconference stations have permanent high priority, produces no degradation in conference quality, but increases the asynchronous network delay. Finally, the time-adjusted priority system provides a trade-off between the first two schemes. The quality of video conferencing is better than in the first scheme, while the asynchronous network delays are shorter than in the second scheme.

Present optical network technology can support the broadband integrated services digital networks (BISDN) standard and has become the key network for multimedia applications. The two B channels of the ISDN basic access provide $2 \times 64$ kbps of composite bandwidth. Conferences can use part of this capacity for wideband speech, saving the remainder for purposes such as control, meeting data, and compressed video. BISDN [2] networks are in either synchronous transfer mode (STM) or asynchronous transfer mode (ATM) to handle both constant and variable bit rate traffic applications. STM provides fixed bandwidth channels and therefore is not flexible enough to handle the different types of traffic typical in multimedia applications.

## New Words

corruption [kəˈrʌpʃən] n. 腐败,堕落,恶化;贪污,贿赂
lightweight [ˈlaitweit] adj. 不重要的,无足轻重的(人)
protocol [ˈprəutəkɔl] n. (外交的)议定书;会谈记录[备忘录]
entail [inˈteil] vt. 把(疾病等)遗传给,把……遗留给(on, upon)
videoconference [ˌvidiəuˈkɔnfərəns] n. 视频会议

## Phrases & Expressions

trade-off    权衡,折中

## Technical Terms

multicasting [ˈmʌltikɑːstiŋ] n. [电]多广播
latency [ˈleitənsi] n. [计]传输时间,传输延迟时间
jitter [ˈdʒitə] vi. [计]跳动,抖动
priority [praiˈɔriti] n. 优先权,优先;[计]优先级
synchronous [ˈsiŋkrənəs] adj. 同步
asynchronous [eiˈsiŋkrənəs] adj. 不同时的;[电]异步的
token ring    令牌环

## Notes

1. Traditional networks are used to provide error-free transmission. However, most multimedia applications can tolerate errors in transmission due to corruption or packet loss without retransmission or correction.

传统网络提供无差错传输。而在大多数多媒体应用中,由于没有重传或者更正而造成的传输失败或丢包错误是可以接受的。

2. BISDN:Broadband Integrated Services Digital Network,宽带综合业务数字网,是在ISDN的基础上发展起来的,主要特征是以同步转移模式(STM)和异步转移模式(ATM)兼容方式,在同一网络中支持范围广泛的图像、声音和数据的应用。ATM不但能把图像、数据、语音等各种业务都综合到一个网内,还有可实现多媒体通信和带宽动态分配的优点。

## Exercises

### 1. Keywords

In the article, there are some important words which are the soul of this paper. After reading this paper, we can find some words to stand for this article. Now please find out key words of this paper.

### 2. Summary

After reading this paper, please write a summary about 3G.

# 科技英语知识 11：科技英语翻译标准

科技文献作为现代科技的载体之一，主要是述说事理、描写现象、推导公式和论证规律。因此，科技文献的主要特点是行文规范、表述自然、用词考究。在科技文献中注重事实的陈述，不添加作者个人的感情色彩，因此在翻译时应注重逻辑思维清楚、语言描述规范、含义表达准确。

翻译时主要要求遵循的是"信、达、雅"三原则，该原则通常被认为是指导翻译实践、鉴定翻译成败和评价译文优劣的经典标准。广义上讲，三原则适用于一切翻译活动，在科技文献翻译时，其翻译标准在此基础上根据具体要求会有所侧重和扩展，即通常所说的"信、达、雅、捷"。

1. "信"，即忠实于原文，也就是完整地再现原文的内涵和思想。科学的灵魂是"真"，科技翻译的灵魂是"准确"。译文应避免漏译或错译，忠实正确地表达原文的内容，既不歪曲，也不随意增减。同时，在表达上要保持原作的风格和文体。

2. "达"，即通顺，就是通过流畅、逻辑的语言表达原文。只追求忠实而不顾通顺，将使译文枯燥无味，也不是标准的翻译。在原文"蓝本"的基础上得到"通顺"的译文，需要译者对行文方式和表达手法反复推敲、仔细权衡，在比较选择中寻找并确定最佳的表现手段，尽可能完美地复现原文的全部内容。

3. "雅"，即在重视和通顺的基础上还应注意"文采"。翻译是要讲修辞的，以使译文逻辑严谨流畅、语言优美易懂，否则就会出现"言之无文"而使读者感觉乏味。在译文中适当增加一些承上启下的连接词，灵活使用汉语中的俗语或者成语等将使文章更加符合汉语标准且更显文采。例如，将"It is well known that ..."译为"众所周知……"，将"As the name suggests, ..."译为"顾名思义，……"。

4. "捷"，即希望有较高的翻译速度。通常来说，速度越快，获取的信息量越大，效率也就越高。

# Unit 13  Digital Multimedia Systems

## Lesson 35  Basic Concepts in Digital Multimedia Systems

Humans communicate using a variety of senses and capabilities, especially in face-to-face situations. We should aim to emulate the bandwidth, fidelity, and effectiveness possible in those situations when we develop interactive multimedia computing systems, especially as we move from analog to digital processing environments.

That movement, a part of the evolution of information technology since the early days of computing, gained momentum with the widespread use of compact discs, which demonstrated the accurate reproduction and superb quality of digital audio. Bilevel (black-and-white) image handling, especially facsimile, has demonstrated the potential for rapid communication of documents, changing in a few years the way organizations operate.

Methods for managing computer graphics, color images, and motion video will lead to even greater changes. When fully digital multimedia computing systems are readily available, we will have powerful tools for improving human-human collaboration and human-computer symbiosis.

Televisions, CD players, telephones, and home computers will evolve and be combined, yielding systems with stereo speakers, high-resolution color displays, megabytes of RAM, fast processors for video and audio, fiber-optic network connections, hundreds of megabytes of disk capacity, CD-ROM drives, and flexible input devices, including stereo microphones, pointing devices, and text entry units. True programming of video will be possible for personalized presentations.

High-resolution images, high-fidelity audio, nicely typeset text, and high-definition video will be available on demand, as versatile alternatives to conventional photographic, audio, newspaper, and television services. Home shopping, cottage industries, delivery of professional services, supplemental adult and child education, surrogate travel to real or artificial sites, video mail and conferencing, and diverse modes of entertainment will be supported.

Many areas of computer science and electrical engineering are aiding these developments. Fast processors, high speed networks, large-capacity storage devices, new algorithms and data structures, graphics systems, innovative methods for human-computer interaction, real-time operating systems, object oriented programming, information storage and retrieval, hypertext and hypermedia, languages for scripting[1], parallel processing methods, and complex architectures for distributed systems all are involved.

To understand interactive digital multimedia computing systems, it is necessary to see how relevant aspects of these fields relate.

First, we need known background regarding developments in interactive videodiscs. Then, we also known digital storage media, including optical, magnetic, and network options, which allow digital multimedia to be preserved, shared, and distributed. We need learn the characteristics of audio and video and their digital representations. Because these media are so demanding of space and channel bandwidth, we review compression methods.

Technology is not all that is necessary. Vendors must follow standards discussed in their own section to ensure that the economics and usability of digital multimedia help this industry grow. Building on existing de facto standards and an emerging suite of international standards, digital multimedia systems (for example, from Intel, Commodore, and Philips) are already available, and the future for digital multimedia in general looks bright.

### Interactive videodiscs

Computer handling of large quantities of audio and video information became possible with the advent of the videodisc in the late 1970s. Each side of these optical discs can hold 54 000 images, or 30 minutes of motion video if the images are played in sequence at the standard rate of 30 frames per second and they run concurrently with 30 minutes of stereo sound. All recorded in an analog format. Although seek time is on the order of a second, the random access capability allows computers to control playback in interactive videodisc systems. Videodisc output usually goes directly to a monitor; with additional boards the computer system can overlay text or graphics on the video output, or even digitize the video signal as it is received.

Preparation of videodisc applications is typically a relatively expensive process, requiring a team for design, video and audio production, graphic art, programming, project management, and content specialist duties. While mastering and replication cost several thousand dollars, completes projects may cost 100 000 per disk. Recordable videodiscs are available but not common, so preparing videodiscs is essentially a publication process. When interactive videodiscs are coupled with high quality software and a good user interface, powerful educational experiences for thousands of people can result.

## New Words

bilevel [bai'levl] *n.* [医]双相的
facsimile [fæk'siməli:] *n.* 无线电传真,传真电报
supplemental [ˌsʌpli'mentl] *adj.* 补足的,追加的
surrogate ['sʌrəgit] *n.* 代理人,代表,委员
diverse [dai'vəːs] *adj.* 不同的;多变化的
retrieval [ri'triːvəl] *n.* (可)取(收)回,恢复,挽回;挽救
interactive [ˌintər'æktiv] *adj.* 相互作用

facto ['fæktəu] *n.* 事实上,实际上
innovative ['inəuveitiv] *adj.* 创新的;革新(主义)的
superb [ˌsjuː'pəːb] *adj.* 宏伟的;壮丽的;华美的;[口]极好的,超等的

## Phrases & Expressions

de facto    事实上的

## Technical Terms

fidelity [fi'deliti] *n.* [电]逼真度,忠诚,忠贞,忠实
megabytes    百万(兆)字节
hypertext ['haipəˌtekst] *n.* [计]超文本系统
mastering    母盘
motion video    运动视频
object oriented programming    面向对象编程
parallel processing    并行处理

## Note

1. scripting：脚本,在网页中插入的脚本有 JavaScript 和 VBScript 两种,它们都是解释性的脚本语言,其代码可以直接嵌入 HTML 命令中。JavaScript 和 VBScript 的最大特点是可以很方便地操纵网页上的元素,并与 Web 浏览器交互,同时,JavaScript 和 VBScript 可以捕获客户的操作并做出反应。

## Lesson 36  DVD

The acronym DVD(for digital versatile disc, or digital video disc)refer to a relatively new high-capacity optical storage format that can hold from 4.7 GB to 17 GB, depending on the number of recording layers and disc sides being used. A standard single layered, single-sided DVD can store 4.7 GB of data;a two-layered standard enhances the single-sided layer to 8.5 GB. DVD can be double-sided with a maximum storage of 17 GB per disc. The DVD was initially developed to store the full contents of a standard two-hour movie, but is now also used to store computer data and software. DVD-ROM technology is seen by many as the successor to music CDs, computer CD-ROMs, and prerecorded VHS videotapes that people buy and rent for home viewing-in other words, read-only products. DVD-R is DVD recordable (similar to CD-R). The user can write to the disk only once. DVD-RW is DVD rewritable (similar to CD-RW). The user can erase and rewrite to the disk multiple times. Only one-sided disks can be used for both DVD-R and DVD-RW. Most DVD drives can play both computer and audio CDs, but you can't play DVDs in a CD drive.

**DVD Forum Receives Top Information Technology Industry Award for Creation of Unified Specification for Next Digital Multimedia Era.**

The DVD Forum today announced that it has received the 1997 PC Magazine Award for Technical Excellence in the category of "Standards", in recognition of the Forum's successful development of the DVD-ROM specification. "New standards are particularly important, since they promise to bring higher levels of technology innovation and market compatibility to today's technology users," said Michael J. Miller, editor in chief of PC Magazine. "DVD ROM is a compelling technology that was chosen because it's a familiar format that brings a wealth of new computing, educational, gaming and entertainment possibilities to the user. In the Award citation to DVD-ROM technology, PC Magazine referred to DVD as the format that will "replace the CD-ROM as the primary means of PC content distribution."

Representatives of three companies involved in development of the specification. Hitachi Ltd., Matsushita Electric Industrial Company(Panasonic)and Toshiba Corporation, accepted the award on behalf of the DVD Forum in a ceremony held on November 17 at COMDEX' 97 in Las Vegas. "Products based on specifications defined by the DVD Forum are now shipping in volume to the worldwide computer and consumer electronics markets, and the Award for Technical Excellence adds to the market's validation of the success of the standards process, "said Koji Hase, General Manager of the DVD Products Division at Toshiba Corporation and a founding member of the DVD Forum. "We are extremely pleased to see the work of the DVD Forum recognized as one of the key technical achievements in the personal computer industry, particularly as the Forum expands the scope of its work with a larger global membership in 1998."

"The members of the DVD Forum developed the DVD-ROM specification as the best technical approach and also the best approach for customers in the marketplace," said Sakon Nagasaki, director of the DVD Business Development Office of Matsushita Electric Industrial Co. Ltd. (Panasonic). "Acceptance of the format illustrates how standards-making efforts advance the goals of the entire electronics industry."

In addition to its role in development of the DVD-ROM and DVD-Video standards, the Forum has proposed the format for recordable DVD, known as DVD-R, and rewrite-able DVD, known as DVD-RAM[1], to international standards bodies. Work is also continuing on definition of a DVD Audio specification.

"The mission of the DVD Forum is to define a smooth migration path from CD to DVD technology by working with the widest possible representative group of manufacturers and technology end-users in the converging industries of computers and consumer electronics," said Dr. Yoshita Tsunoda, a member of the Executive Staff of Hitachi, Ltd. and Chairman of the DVD Forum's DVD-RAM Working Group. "The different working groups have already completed definition of three separate DVD technology standards, and we have begun work on developing next generation specifications that will provide compatible products well into the next century."

Recipients of the PC Magazine Awards for Technical Excellence are named by a team of editors, senior contributors and PC Labs personnel after months of evaluation and discussion. PC Magazine, the sponsor of the Technical Excellence Awards, is a 1.175 million circulation magazine published by Ziff-Davis Inc. PC Magazine is published 22 times a year in print, quarterly on CD and continuously on the World Wide Web.

## New Words

acronym ['ækrənim] n. 首字母缩略词
enhance [in'hɑ:ns] vt. 提高,增加;加强
maximum ['mæksiməm] adj. 最大值的,最大量的; n. 最大的量、体积、强度等
prerecord ['pri:ri'kɔ:d] vt. 事先录音
recognition [ˌrekəg'niʃn] n. 认识,识别;承认,认可;褒奖;酬劳
innovation [ˌinə'veiʃn] n. 改革,创新;新观念;新发明;新设施
compelling [kəm'peliŋ] adj. 引人入胜的,扣人心弦的,非常强烈的; v. 强迫,使不得不
　　　　　(compel 的现在分词)
COMDEX COMputer Dealer's EXpo　计算机分销商展览
marketplace ['mɑ:kitpleis] n. 市场,集市;商业界

## Phrases & Expressions

optical storage format　光存储格式

## Technical Terms

VHS Video Home System　家用录像系统
DVD　数字化视频光盘
DVD-R　只读 DVD
DVD-RW　可读写 DVD

## Note

1. DVD-RAM: DVD-Random Access Memory，DVD Forum 于 1996 年制订的光盘规格，限定了可多次写入的 DVD-RAM 存储媒体及 DVD 刻录机所使用的格式。DVD-RAM 自 1998 年起广泛使用于计算机的光驱、刻录机、录像机、手提摄录机等。

# Lesson 37　Compression Methods

Compression is essential if audio, images, and video are to be used in digital multimedia applications. A megabyte of space would be filled by roughly six seconds of CD-quality audio, a single 640×480-pixel color image stored using 24 bits per pixel, or a single frame - 1/30 second - of CIF[1] video. Nevertheless, videodisc applications often have more than 20 minutes of video, perhaps 10 000 slides, and 30 minutes of stereo sound on each laser disc side. And with the tremendous volume of data that will be received each day from planned NASA missions and other scientific ventures, the need for proven compression techniques is obvious.

Happily, there has been a great deal of research and many implementations using software, hardware, or both for a variety of compression methods. Research continues, with further improvements expected using wavelet and other time and space domain schemes. A set of useful articles appears in conference publications by NASA [2] and the IEEE [3] Computer Society, with many good references included in the bibliographies.

Compression of digital data involves computational algorithms that can be implemented in software. Some involve digital versions of signal processing methods, others involve pattern recognition, and still others use statistics or characteristics of particular data types or samples. High-speed implementations involve VLSI chips, such as for audio digital signal processing, discrete cosine transform, or vector quantization approaches.

At the boundary of image processing, computer vision, and graphics is the area of model-based compression. Models of faces can be analyzed, yielding facial motion parameters that can be transmitted at low bit rates and synthesized at a receiver for "talking head" video telephony. Other approaches involve feature detection at the encoder and rendering at the decoder.

Fractals (images that can be described by a set of rules specified with a relatively small number of bits) allow compression of natural scenes where the underlying structure matches this type of model. Very high compression ratios can be achieved, sometimes on the order of 1 000 : 1 (size of the uncompressed form versus the compressed form). However, extensive computation is required for encoding. While decoded images may be acceptable to human judgment, there is usually some quality loss. Nevertheless, several companies—for example, Barnsley Communications — are marketing boards and software for fractal compression and decompression. Commodore has announced plans for software based fractal decompression in their CDTV system.

Figure 37-1 shows a taxonomy of compression approaches. In lossless schemes, the original representation can be perfectly recovered. For text, lossless methods may achieve a 2 : 1 reduction. For bilevel images, 15 : 1 is a good figure. (A new international standard for bilevel image coding, referred to as JBIG [4], improves on CCITT [5] Group 3 and Group 4 approaches for facsimile transmission, and in some situations achieves more than 50 : 1 compression).

These approaches are also called noiseless - because they do not add noise to the signal - or entropy coding—because they eliminate redundancy through statistical or decomposition techniques. For example, Huffman coding uses fewer bits for more common message symbols, and run-length encoding replaces strings of the same symbol with a count/symbol pair.

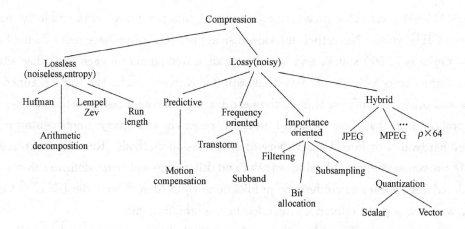

Figure 37-1  Selected compression approaches

The other approach, lossy compression, involves encoding into a form that takes up a relatively small amount of space, but which can be decoded to yield a representation that humans find similar to the original [6].

● Lossy compression: Lossy or noisy compression may add artifacts that can be perceived. Careful study of the human visual system has focused attention on approaches that cause little perceived loss in quality, but achieve high compression ratios.

● Prediction: Predictive approaches like ADPCM involve predicting subsequent values by observing previous ones, and transmitting only the usually small differences between actual and predicted data. An example involves motion compensation. Successive frames in a video sequence are often quite similar or have blocks of pixels shifted from one frame to the next—for example, as the camera pans or a person moves. Although it is computationally expensive to analyze images and yield motion vectors, parallel computers or neural networks can help with the processing.

● Frequency-oriented compression: Subband coding can exploit the fact that humans have different sensitivities to various spatial and temporal frequency combinations. The idea is to separate (for example, using a series of filters) the different frequency combinations, and then to code with greater fidelity the frequencies that humans pay particular attention to. Without subband coding, all frequency combinations would be coded identically, so the technique achieves high perceived quality with fewer total bits.

Another approach relating to humans' handling of frequency is transform coding. This usually involves spatial frequencies, as in single images. The most common approach applies the discrete cosine transform, which is related to the fast Fourier transform. Lower spatial frequencies must be carefully coded, while higher frequencies need less detailed coding. If we think of a block (say, an $8\times8$-pixel section) of a two-dimensional image as a square with rows and columns numbered from the top-left corner, then the DCT of that block will also be a similarly numbered square [7]. Consider a zigzag sequencing of the values in transform space, starting at the top-left corner and covering the nearest cells first. Run-length encoding and coarse quantification of cells later in the sequence both lead to good compression. The encoder applies DCT in the forward direction, and the decoder uses an inverse mapping from transform to image space.

• Importance-oriented compression: Other characteristics of images besides frequency are used as the basis for compression. The principle is to consider as more important those parts of an image that humans are better attuned to. An example of this approach is to filter images, getting rid of details that cannot be perceived, as in the low-pass filtering done for real-time video with DVI systems. Another technique is to allocate more bits to encode important parts of an image, such as where edges occur, than to encode large homogeneous regions, such as those depicting clouds.

Color lookup table use, as in CD-I and DVI, applies the principle of indirection. Instead of letting the bits that describe a pixel refer to a location in color space, the bits identify a table location, and the table entry refers to color space. Color spaces often cover a palette of size $2^{24}$, which means 24 bits are needed. On the other hand, lookup table size may be only 256 ($2^8$). The reduction is 24 to 8 bits per pixel. The challenge is to select for each lookup table the most important colors to be accessed by the display processor.

Subsampling, also based on characteristics of human vision. It involves using fewer bits for chrominance than luminance. Interpolation, which can be carried out in hardware, results in a full but approximate reconstruction of the original. We can think of this process as that of taking one matrix and generating from it another matrix four or 16 times larger — by interpolating values horizontally, vertically, and diagonally. Related to interpolation is line doubling, used in some DVI systems to go from the 256 lines that result from video decompression to 512 lines.

Importance also relates to patterns in an image representation. Clearly, higher level descriptions where symbols refer to large structures can take much less space than raster forms. In coding theory, this translates into the fact that vector quantization can lead to higher compression than scalar quantization. Scalar quantization is often just called quantization, and was discussed in an earlier section in connection with pulse code modulation and audio encoding. It takes values and maps them into a fixed number of bits.

Vector quantization, on the other hand, usually takes two-dimensional vectors of values- for example, $4 \times 4$, and maps them into a code symbol. Thus, code books are developed for images, recording the most important vectors, and all data vectors are mapped to the nearest code-book entry, minimizing mean square error. Decoding involves fast table lookup to replace coded entries with vectors from the code book. Encoding usually takes a good deal of computation, so near optimal code books can be developed.

• Hybrid coding: Various compression approaches can be combined, for example, DCT and differential pulse code modulation, subband coding and DCT, or differential pulse code modulation and vector quantization. Generally, subband coding is coupled with vector quantization. Systems and standards for video compression often apply motion compensation for temporal compression, transform coding for spatial compression, and Huffman or arithmetic coding for statistical compression.

## New Words

  megabyte ['megəbait] n. 兆字节；百万字节
  slide [slaid] n. 滑，滑道，幻灯片
  stereo ['stiəriəu] adj. 立体的，立体感觉的；[电]立体
  tremendous [tri'mendəs] adj. 巨大的，非常的，可怕的
  bibliography [ˌbibli'ɔgrəfi] n. 参考书目；[计]书目，文献目录
  fractal ['fræktəl] n. 分形
  taxonomy [tæk'sɔnəmi] n. 分类学，分类系统
  eliminate [i'limineit] vt. 除去，排除，剔除，消除
  redundancy [ri'dʌndənsi] n. 冗长，多余，冗余位，冗余码；[计]冗余
  artifact ['ɑːtifækt] n. 人工制品，人工产物
  spatial ['speiʃəl] adj. 空间的；立体的，三维的
  temporal ['tempərəl] adj. 时间的，暂时的
  zigzag ['zigzæg] adj. Z字形的，之字形的，锯齿形的
  palette ['pælit] n. 调色板，颜料；[计]调色板
  diagonal [dai'ægənl] adj. 对角线的，斜的，斜纹的

## Phrases & Expressions

  be coupled with 和……联合，结合

## Technical Terms

  pixel ['piksəl] n. [计]像素
  wavelet ['weivlit] n. 小波
  subsampling 二次抽样
  single frame 单帧；一幅画面
  laser disc 激光影碟
  chrominance ['krəuminəns] n. 色品，彩色信号，色度；[计]色度
  luminance ['ljuːminəns] n. 亮度；[计]亮度
  interpolation [inˌtəːpəu'leiʃən] n. 插入(值)，内插；插值法，内插(推)法，插值法
  raster ['ræstə] n. 光栅；[电]试映图
  pattern recognition 模式识别
  DCT(Discrete Cosine Transform) 离散余弦变换
  vector quantization 矢量量化

CDTV （Conventional-Definition Television） 普通清晰度电视
bi-level image  二值图像，黑白图像
facsimile transmission  传真发送
entropy coding  熵编码
lossy compression  有损压缩
ADPCM  自适应音频脉冲编码
motion compensation  动态补偿，运动补偿
neural network  神经网络
fast Fourier transform  快速傅里叶变换
DVI  交互式数字视频系统
CD-I  交互式CD
hybrid coding  混合编码法

# Notes

1. CIF：Common Intermediate Format，常用的标准化图像格式。H.323 协议族中规定了视频采集设备的标准采集分辨率。CIF=352 像素×288 像素。

2. NASA：National Aeronautics and Space Administration，美国国家航空航天局，又称美国宇航局或美国太空总署，是美国负责国家太空计划的一个政府部门。

3. IEEE：Institute of Electrical and Electronics Engineers，美国电气与电子工程师学会，是一个建立于 1963 年 1 月 1 日的国际性电子技术与信息科学工程师协会，也是世界上最大的专业技术组织之一，拥有来自 175 个国家的 36 万名会员。它定位在"科学和教育，并直接面向电子电气工程、通信、计算机工程、计算机科学理论和原理研究的组织，以及相关工程分支的艺术和科学"。为了实现这一目标，IEEE 担任多个科学期刊和会议组织者的角色。它也是一个广泛的工业标准开发者，主要涉及领域包括电能、能源、生物技术和保健、信息技术、信息安全、通信、消费电子、运输、航天技术和纳米技术。

4. JBIG：Joint Bi-level Image experts Group，二值图像压缩的国际标准。二值图像的每个像素只用一位表示。JBIG 技术已相当完善，并且广泛应用于传真、印刷等领域。但是 JBIG 只考虑单幅二值图像压缩，并没有考虑动态二值视频帧间的相关性。用 JBIG 直接进行二值动态视频的压缩，数据量仍然很大。

5. CCITT：International Telephone and Telegraph Consultative Committee，国际电报电话咨询委员会。它是国际电信联盟(ITU)的常设机构之一，主要职责是研究电信的新技术、新业务和资费等问题，并对这类问题通过建议使全世界的电信标准化。

6. The other approach, lossy compression, involves encoding into a form that takes up a relatively small amount of space, but which can be decoded to yield a representation that humans find similar to the original.

另一种方法，即有损压缩，是将编码转换成一种占用相对较少空间的形式，其质量与原始数据质量相比没有明显差别。

7. If we think of a block (say, an 8×8-pixel section) of a two-dimensional image as a

square with rows and columns numbered from the top-left corner, then the DCT of that block will also be a similarly numbered square.

如果我们考虑二维图像的一个块（也就是说，一个 8 像素×8 像素部分）作为一个从左上角进行行和列编号的正方形，那么该块的 DCT 也将是一个同样编号的正方形。

# Exercises

1. Keywords

In the article, there are some important words which are the soul of this paper. After reading this paper, we can find some words to stand for this article. Now please find out key words of this paper.

2. Summary

After reading this paper, please write a summarize about compression methods.

# 科技英语知识 12：科技论文的结构与写作

科技论文是科技人员介绍有关研究成果的文章，从结构上通常分为标题（title）、作者（author）、摘要（abstract）、关键词（keywords）、正文部分（通常由引言 introduction、主体 body 和结论 conclusion 构成）以及致谢（acknowledgment）和参考文献（reference）几个部分。下面主要介绍论文正文部分的写作。

论文的开头，即引言部分，主要介绍与论文有关的背景知识及现存的问题，从而引出论文的写作目的和主题。

其常用的英语句型有：

The experiments (research) on ... was carried by ...

Recent experiments by ... have suggested that ...

The previous work on ... has indicated that ...

The paper is divided into five major sections as follows ...

论文的主体部分篇幅大、内容多，是主体思想的展开和论述部分，选材上要围绕主题，段落划分既要结构严谨，又要保证全文的整体性和连贯性。作者可根据需要加小标题，将主题内容分为几个部分进行论述。英文写作通常把每段的主体句（topic sentence）放在段落的第一句，全段围绕主体句论述，必要的例子和对试验数据的分析通常是不可缺少的。

常用的句型有：

The problem is chiefly concerned with the nature (effect, activity) of ... based on the study of ...

The core of the problem is the interaction (origin, connection) of ...

It is described as follows ...

The present study was made with a view to show (demonstrate, determine) ...

Studies of these effects cover various aspects of ...

结论是对研究或试验结果（仿真结果）的分析，要阐明作者的观点，指出争议的问题，得出最后的结论。

常用的句型有：

On the basis of ..., the following conclusion can be made ...

From ..., we now conclude (sum up) that ...

We have demonstrated in this paper ...

The results of the experiment (simulation) indicate (show) ...

Finally, a summary is given of ...

此外，为了保证论文内部应有的内在联系，文中应适当使用一些连接词语，常用的有以下几种。

并列：and, also, in addition, as well as；

转折: although, even though, but, on the other hand, otherwise, while, yet, in spite of;
承接: first of all, secondly, thirdly, furthermore, moreover, besides, finally;
条件: if, whether, unless, on condition that, so long as;
举例: for example, for instance, as an illustration, such as;
原因: for, because, for that reason, as, since, in order to;
结果: as a result of, hence, so, consequently, thus, therefore, then;
总结: in conclusion, to sum up, to summarize, to conclude。

# Unit 14　Electric Systems

## Lesson 38　Personal Computer

A personal computer (PC) is any general-purpose computer whose size, capabilities, and original sales price make it useful for individuals, and which is intended to be operated directly by an end user, with no intervening computer operator.

As of 2009, a PC may be a desktop computer, a laptop computer or a tablet computer. The most common operating systems for personal computers are Microsoft Windows, Mac OS and Linux, while the most common microprocessors are x86-compatible CPUs, ARM architecture CPUs and PowerPC CPUs. Software applications for personal computers include word processing, spreadsheets, databases, Web browsers and E-mail clients, games, and myriad personal productivity and special-purpose software. Modern personal computers often have high-speed or dial-up connections to the Internet, allowing access to the World Wide Web and a wide range of other resources.

A PC may be a home computer, or may be found in an office, often connected to a local area network (LAN). This is in contrast to the batch processing or time-sharing models which allowed large expensive systems to be used by many people, usually at the same time, or large data processing systems which required a full-time staff to operate efficiently [1].

While early PC owners usually had to write their own programs to do anything useful with the machines, today's users have access to a wide range of commercial and non-commercial software which is provided in ready-to-run form.

The capabilities of the PC have changed greatly since the introduction of electronic computers. By the early 1970s, people in academic or research institutions had the opportunity for single-person use of a computer system in interactive mode for extended durations, although these systems would still have been too expensive to be owned by a single person. The introduction of the microprocessor, a single chip with all the circuitry that formerly occupied large cabinets, led to the proliferation of personal computers after about 1975. Early personal computers — generally called microcomputers — were sold often in Electronic kit form and in limited volumes, and were of interest mostly to hobbyists and technicians. Minimal programming was done by toggle switches, and output was provided by front panel indicators. Practical use required peripherals such as keyboards, computer terminals, disk drives, and printers. Unlike other hobbyist computers of its day, which were sold as kits; in 1976 Steve Jobs and Steve Wozniak sold the Apple I was a fully assembled circuit board containing about 30 chips. Such that by 1977 Apple Computers introduced the Apple II, as the world's first personal computer.

By 1977, mass-market pre-assembled computers allowed a wider range of people to use computers, focusing more on software applications and less on development of the processor hardware.

Throughout the late 1970s and into the 1980s, computers were developed for household use, offering personal productivity, programming and games. Somewhat larger and more expensive systems (although still low-cost compared with minicomputers and mainframes) were aimed for office and small business use. Workstations are characterized by high-performance processors and graphics displays, with large local disk storage, networking capability, and running under a multitasking operating system. Workstations are still used for tasks such as computer-aided design, drafting and modelling, computation-intensive scientific and engineering calculations, image processing, architectural modelling, and computer graphics for animation and motion picture visual effects.

Eventually the market segments lost any technical distinction; business computers acquired color graphics capability and sound, and home computers and game systems users used the same processors and operating systems as office workers. Mass-market computers had graphics capabilities and memory comparable to dedicated workstations of a few years before. Even local area networking, originally a way to allow business computers to share expensive mass storage and peripherals, became a standard feature of the personal computers used at home.

Though a PC comes in many different form factors, a typical personal computer consists of a case or chassis in a tower shape (desktop) and the following parts (Figure 38-1):

① scanner;
② CPU (microprocessor);
③ primary storage (RAM);
④ expansion cards (graphics cards, etc.);
⑤ power supply;
⑥ optical disc drive;
⑦ secondary storage (Hard disk);
⑧ motherboard;
⑨ speakers;
⑩ monitor;
⑪ system software;
⑫ application software;
⑬ keyboard;
⑭ mouse;
⑮ external hard disk;
⑯ printer.

These components can usually be put together with little knowledge to build a computer. The motherboard is a main part of a computer that connects all devices together. The memory card(s), graphics card and processor are mounted directly onto the motherboard (the processor in a socket and the memory and graphics cards in expansion slots). The mass storage is connected to it with cables and can be installed in the computer case or in a separate case. This is the same for the keyboard and mouse, except that they are external and connect to the I/O panel on the back of the computer. The monitor is also connected to the I/O panel, either through an onboard port on the motherboard, or a port on the graphics card.

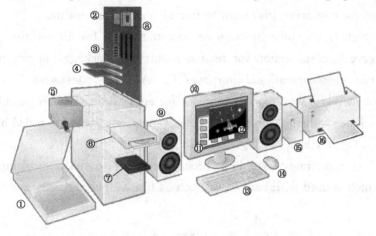

Figure 38-1  Computer consists

Several functions (implemented by chipsets) can be integrated into the motherboard, typically USB and network, but also graphics and sound. Even if these are present, a separate card can be added if what is available isn't sufficient. The graphics and sound card can have a break out box to keep the analog parts away from the electromagnetic radiation inside the computer case. For really large amounts of data, a tape drive can be used or (extra) hard disks can be put together in an external case.

The hardware capabilities of personal computers can sometimes be extended by the addition of expansion cards connected via an expansion bus. Some standard peripheral buses often used for adding expansion cards in personal computers as of 2005 are PCI, AGP (a high-speed PCI bus dedicated to graphics adapters), and PCI Express. Most personal computers as of 2005 have multiple physical PCI expansion slots. Many also include an AGP bus and expansion slot or a PCI Express bus and one or more expansion slots, but few PCs contain both buses.

Computer software, or just software is a general term used to describe a collection of computer programs, procedures and documentation that perform some tasks on a computer system.

The term includes:
• Application software such as word processors which perform productive tasks for users.
• Firmware which is software programmed resident to electrically programmable memory devices on board mainboards or other types of integrated hardware carriers.
• Middleware which controls and coordinates distributed systems.
• System software such as operating systems, which interface with hardware to provide the necessary services for application software.

Software testing is a domain independent of development and programming. It consists of various methods to test and declare a software product fit before it can be launched for use by either an individual or a group. Many tests on functionality, performance and appearance are conducted by modern testers with various tools such as QTP, Load runner, Black box testing etc. to edit a checklist of requirements against the developed code [2]. ISTQB is a certification that is in demand for engineers who want to pursue a career in testing.

Testware which is an umbrella term or container term for all utilities and application software that serve in combination for testing a software package but not necessarily may optionally contribute to operational purposes[3]. As such, testware is not a standing configuration but merely a working environment for application software or subsets thereof.

Software includes websites, programs, video games, etc. that are coded by programming languages like C, C++, etc.

"Software" is sometimes used in a broader context to mean anything which is not hardware but which is used with hardware, such as film, tapes and records.

## New Words

compatible [kəm'pætəbl] adj. [计]兼容的;符合的;能共存的;相容的
size [saiz] n. 大小,尺寸,面积,容量,数量,规模;实力,范围
capability [ˌkeipə'biliti] n. 能力,才能,手腕;性能,容量,功率,生产率
intervene [ˌintə'viːn] vi. 干涉,干预,插进,介入;调停,调解
commercial [kə'məːʃəl] adj. 商业的;商品化的;商用的;质量低劣的
kit [kit] n. [计]成套部件;成套零件;装备,工具箱
hobbyist ['hɔbiist] n. [计]业余爱好者;有业余癖好者
toggle ['tɔgl] n. 触发器;解发器反复电路;切换
animation [ˌæni'meiʃən] n. [计]动画;活泼,生气,激励,卡通制作
umbrella [ʌm'brelə] n. 伞,雨伞; adj. 伞的,包罗万象的,综合的
orbit ['ɔːbit] n. 轨道,常轨; vt. 绕……轨道而行

## Phrases & Expressions

intend to    打算(做)……,想要(做)……
as of    在……时;到……时为止;从……时起;截至
in contrast to    和……形成对比(对照)
full-time    专任的,全职的
ready-to-run    即时可用的,可即时使用的
access to    接近或使用的机会
aim for    力争……;针对
come in    进来,到达终点,流行起来,当选

either … or …　或者……或者……；不是……就是……
even if　即使,纵然
keep away from　远离,回避

## Technical Terms

database ['deitəbeis] n. 数据库,基本数据
panel ['pænl] n. [电]配电盘,仪表盘；[机]控制板,操纵盘(台)
indicator ['indikeitə] n. 指示者,指示物；[机]指示器,指针,[电]目视仪
peripheral [pə'rifərəl] n. (计算机的)外围设备,周边设备(如显示装置、打印机等)
mainframe ['meinfreim] n. [计]主机,大型机
motherboard ['mʌðəbɔːd] n. 底板,母板；[计]母板,主板
desktop computer　台式计算机
laptop computer　便携式计算机,膝上计算机
tablet computer　平板计算机
software application　应用软件
word process　文字处理
spreadsheet　电子表格
Web browser　网络浏览器
E-mail client　电子邮件客户端
dial-up　拨号
World Wide Web　全球信息网,万维网
local area network　局域网
batch process　批处理
time-share　[计]分时,时间共享；[电]分时
pre-assemble　预组装,预先安装
computer-aided design　计算机辅助设计
optical disc drive　光盘驱动器
USB　通用串行总线
expansion slot　扩展槽
testware　测试工具,测试件
PCI (Peripheral Component Interconnect)　周边元器件扩展接口
AGP(Accelerated Graphics Port)　加速图像处理端口
ISTQB(International Software Testing Qualification Board)　国际软件测试工程师认证

## Notes

1. This is in contrast to the batch processing or time-sharing models which allowed large expensive systems to be used by many people, usually at the same time, or large data processing systems which required a full-time staff to operate efficiently.

个人计算机操作模式与在同一时间允许多人使用巨大昂贵系统的分时操作模式或需要专职人员高效操作大量数据处理系统的批处理操作模式相反。

2. Many tests on functionality, performance and appearance are conducted by modern testers with various tools such as QTP, Load runner, Black box testing etc., to edit a checklist of requirements against the developed code.

许多功能、性能和外观的测试由现代测试仪器用不同的测试工具进行，这些测试工具有QTP、Load runner 和黑盒子测试等。测试仪器用这些测试工具编辑一个针对开发代码的检查清单。

3. Testware which is an umbrella term or container term for all utilities and application software that serve in combination for testing a software package but not necessarily may optionally contribute to operational purposes.

测试件是对所有实用工具和应用软件的综合概念或包容性概念，这些工具和软件主要指服务于测试软件包而不是专用于操作目的的结合体。

## Lesson 39  Advanced Automated Fingerprint Identification System

In 1982, National Police Agency of Japan(NPA) developed and introduced Automated Fingerprint Identification System(AFLS), which computerized the process of extraction of minutia and the matching-process of minutia from the NPA's database. This was the first system in the world which realized automated latent fingerprint matching.

In 1997, NPA made AFIS advanced by introducing on-line terminals. This made possible to make inquiry from many remote sites such as Prefecture Police Headquarters (PPH) and Police Stations (PS). Thus we could drastically shorten the inquiry time.

In Japan there are 47 prefectures [1], and every prefecture has its own police headquarter. Police executes criminal investigation, public security, traffic regulation and others related public safety. PPH treats fingerprints in its judicial area.

On the other hand, NPA is one of the government organizations and deals with the nation-wide policy related activities. That's why, in order to identify fingerprint, NPA has researched and developed AFIS from the nation-wide standpoint of view.

Outline of inquiry of fingerprint is shown in Figure 39-1.

● Inquiry of latent fingerprint: PPH uses latent fingerprint to identify suspects by making inquiry into the NPA's rolled fingerprints database.

● Inquiry of rolled fingerprint: PS uses rolled fingerprint to confirm criminal records of arrested suspects by making inquiry into the NPA's rolled fingerprints database and criminal records database at the same time.

**AFIS**

The evolution of image-processing technology made it possible to extract minutiae automatically. And the development of the matching algorithm made it possible to identify fingerprint from the NPA's fingerprints database.

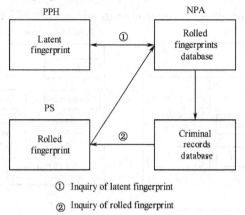

① Inquiry of latent fingerprint
② Inquiry of rolled fingerprint

Figure 39-1  Outline of inquiry of fingerprint

Consequently, in 1982, NPA established the Fingerprint Identification Center which belongs to Identification Division of NPA. There we has stored all the fingerprints.

Identification process of fingerprint in AFIS is shown in Figure 39-2.

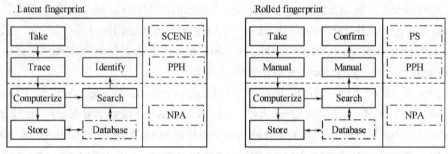

Figure 39-2  Identification process of fingerprint in AFIS

- Process of latent fingerprint: Latent fingerprint is taken at the scene of a crime and then mailed to Identification Division of PPH. It is traced on tracing-paper, and then mailed to Fingerprint Identification Center. There, it is computerized and stored into the latent fingerprints database. At the same time, we make inquiry into the rolled fingerprints database and search it, and then respond to PPH. It takes about 1 week from taking to identification because the process is computerized only at Fingerprint Identification Center.

- Process of rolled fingerprint: Rolled ten fingerprints cards are taken in ink at PS and then mailed to Fingerprint Identification Center via Identification Division of PPH. There, they are computerized and stored into the rolled fingerprints database. At the same time, we make inquiry into its database. And we search it and the criminal records database at the same time, and then respond to PS. It takes about 1 month from taking to configuration because of the above.

### Advanced AFIS

In 1997, NPA introduced on-line remote terminals which are composed of Latent

Fingerprint Inquiry Terminal and Live Scanner. Latent Fingerprint Inquiry Terminal has the function to extract minutiae from tracing paper. Live Scanner has the function to scan fingerprint. Identification process of fingerprint in advanced AFIS is shown in Figure 39-3.

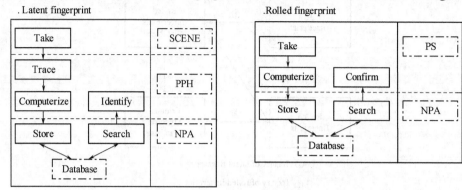

Figure 39-3  Identification process of fingerprint in advanced AFIS

● Process of latent fingerprint: Latent fingerprint is taken at the scene of a crime and then mailed to Identification Division of PPH. There, it is scanned and computerized by Latent Fingerprint Inquiry Terminal and then sent to Fingerprint Identification Center by the satellite network. Afterward it is stored into the latent fingerprints database. And at PPH we make inquiry into the rolled fingerprints database in Fingerprint Identification Center. There, we search the database and respond to PPH. It takes about 1 day from taking to identification because we can make inquiry form remote terminals.

● Process of rolled fingerprint: Rolled ten fingerprints cards are taken at PS, and then they are scanned and computerized by Live Scanner. They are sent to Fingerprint Identification Center by the digital network and satellite network. There, they are stored into the rolled fingerprints database. And at PS we make inquiry into rolled fingerprints database in Fingerprint Identification Center. There we search its database and the criminal records database at the same time, and then respond at PS. It takes about 1 day from taking to confirmation because of the above.

Components

There are 2 components in advanced AFIS. They are Latent Fingerprint Inquiry Terminal and Live Scanner, which NPA has researched and developed.

Latent Fingerprint Inquiry Terminal's, The function is as follows:

(1) scanning fingerprint by image-processing;

(2) clarifying the unclear part of fingerprint;

(3) extracting some minutiae that the ridge part starts, ends and branches automatically.

The function of Live Scanner is scanning rolled fingerprint by image-processing. And we do not use ink and therefore there is the effect to reduce the resistance of arrested suspects for taking rolled fingerprint.

● Principle: Fingerprints has two parts of the type. One is the part of ridge; the other is the part of ditch. When the finger is placed on the surface of the device, the ditch part is not touched and all the light reflects. On the other hand, the ridge part is touched and all the light does not reflects. The device captures the quantity of the light. Thus, the ditch part is large and the ridge part is small in intensity.

### Network

NPA has established the satellite network connecting PPH with Fingerprint Identification Center nation-widely by 1995. The satellite that NPA uses is communications satellite of the private enterprise, which we call Superbird-A.

Bit-rate(1.5 Mbps) is so speedy that it takes less than time to send fingerprint. And NPA always uses 1 channel for advanced AFIS exclusively. The example of distributed frequency is shown in Figure 39-4.

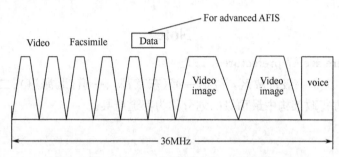

Figure 39-4    Example of distributed frequency

## New Words

computerize [kəm'pjuːtəraiz] vt. (给……)装备电子计算机,(使)电子计算机化
extraction [iks'trækʃən] n. 提取,抽出,拔出,提炼,脱模
minutia [mai'njuːʃiə] n. 细节,小节
inquiry [in'kwaiəri] n. 查询,询问,打听;质询;调查;探究
remote [ri'məut] adj. 遥远的(指时间或地点);偏僻的
drastically ['dræstikli] adv. 激烈地,彻底地
afterward ['ɑːftəwəd] adv. 其后;继后;后来
clarify ['klærifai] vt. 澄清;讲清楚,阐明;使(头脑等)变清楚
intensity [in'tensiti] n. 强烈,剧烈;强度,密度,应力,亮度;明暗度
channel ['tʃænl] n. [电]电路,(电缆的)管;[无]波道,信道,频道

## Phrases & Expressions

PPH    辖区警察总局
PS(Police Stations)    派出所,警察局
be taken in    受骗,被接纳
take to    开始,喜欢,沉溺于,走向,照料,求助于,适应
less than    少于,不到

## Technical Terms

latent fingerprint　（犯罪现场等的）隐约指纹,潜指印,汗潜指印
rolled fingerprint　采集指纹
tracing-paper　描图纸,透明纸
satellite network　卫星网络
bit-rate　比特率
ridge [ridʒ] *n.* 纹路,山脊;山脉
ditch [ditʃ] *n.* 沟渠,沟

## Note

1. In Japan there are 47 prefectures, ...

日本全国分为47个一级行政区,包括一都（东京都）、一道（北海道）、二府（大阪府、京都府）、43县,故此地方行政架构中最顶级的划分称为都道府县。

# Lesson 40　System on a Programmable Chip

SOPC Builder is a powerful system development tool for creating systems based on processors, peripherals, and memories. SOPC Builder enables you to define and generate a complete system-on-a-programmable-chip (SOPC) in much less time than using traditional, manual integration methods. SOPC Builder is included in the Quartus II software and is available to all Altera customers. Many designers already know SOPC Builder as the tool for creating systems based on the Nios II processor. However, SOPC Builder is more than a Nios II system builder; it is a general-purpose tool for creating arbitrary SOPC designs that may or may not contain a processor. SOPC Builder automates the task of integrating hardware components into a larger system. Using traditional system-on-chip (SOC) design methods, you had to manually write top-level HDL files that wire together the pieces of the system. Using SOPC Builder, you specify the system components in a graphical user interface (GUI), and SOPC Builder generates the interconnect logic automatically. SOPC Builder outputs HDL files that define all components of the system, and a top-level HDL design file that connects all the components together. SOPC Builder generates both Verilog HDL[1] and VHDL[2] equally, and does not favor one over the other.

In addition to its role as a hardware generation tool, SOPC Builder also serves as the starting point for system simulation and embedded software creation. SOPC Builder provides features to ease writing software and to accelerate system simulation.

An SOPC Builder component is a design module that SOPC Builder recognizes and can automatically integrate into a system. SOPC Builder connects multiple components together to create a top-level HDL file called the system module. SOPC Builder generates system interconnect fabric that contains logic to manage the connectivity of all components in the system.

SOPC Builder components are the building blocks of the system module. SOPC Builder components use Avalon interfaces for the physical connection of components, and you can use SOPC Builder to connect any logical device (either on-chip or off-chip) that has an Avalon interface. There are two different Avalon interfaces:

● The Avalon Memory-Mapped (Avalon-MM) interface uses an address-mapped read/write protocol that enables flexible topologies for connecting master components to read and/or write any slave components.

● The Avalon Streaming (Avalon-ST) interface is a high-speed, unidirectional, system interconnect that enables point-to-point connections between streaming components that send and receive data using source and sink ports.

SOPC Builder components can use either Avalon-MM or Avalon-ST interfaces or both.

Figure 40-1 shows an FPGA design including an SOPC Builder system module and custom logic modules. Custom logic can integrate inside or outside the system module. In this example, the custom design module inside the system module is an SOPC Builder component and communicates with other modules through an Avalon-MM master interface. The custom logic outside of the system module is connected to the system module through a PIO interface. The system interconnect fabric connects all of the SOPC Builder components using the Avalon-MM or Avalon-ST system interconnect as appropriate.

A component can be a logical device that is entirely contained within the system module, such as the processor component shown in Figure 40-1. Alternately, a component can act as an interface to an off-chip device, such as the SDRAM interface component in Figure 40-1. In addition to the Avalon interface[3], a component can have other signals that connect to logic outside the system module. Non-Avalon signals can provide a special purpose interface to the system module, such as the PIO in Figure 40-1. A component can be instantiated more than once per design. Altera and third-party developers provide ready-to-use SOPC Builder components, including:

- Microprocessors, such as the Nios II processor;
- Microcontroller peripherals, such as a scatter-gather DMA controller;
- Timers;
- Serial communication interfaces, such as a UART and a serial peripheral interface (SPI);
- General purpose I/O;
- Digital signal processing (DSP) functions;
- Communications peripherals, such as a 10/100/1000 Ethernet MAC;
- Interfaces to off-chip devices:

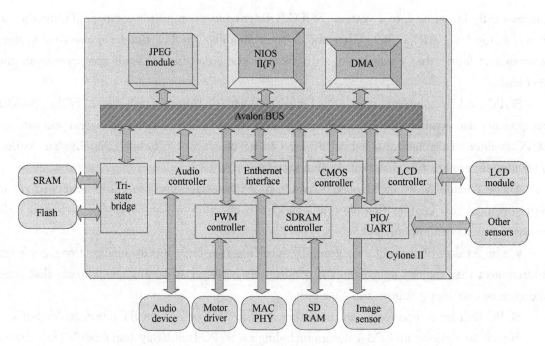

Figure 40-1 Example of a FPGA with a system module generated by SOPC builder

- Memory controllers;
- Buses and bridges;
- Application-specific standard products (ASSP);
- Application-specific integrated circuits (ASIC);
- Processors.

## New Words

arbitrary ['ɑːbitrəri] adj. 任意的, 随意的
embed [im'bed] vt. 使插入, 使嵌入, 使深留脑中; [计]嵌入
fabric ['fæbrik] n. 构造, 结构, 组织; 体制
unidirectional [ˌjuːniˈdirekʃənəl] adj. 单向的, 单向性的
alternately [ɔːlˈtəːnitli] adv. 交替地, 间隔地
instantiate [inˈstænʃieit] vt. 用具体例证说明
certification [ˌsəːtifiˈkeiʃən] n. 证明, 保证, 鉴定, 证明书

## Technical Terms

PIO interface  并行输出入接口
plug-and-play  即插即用
top-level  [计]顶层
IP(intellectual property)  知识产权

## Notes

1. Verilog HDL：是一种硬件描述语言（Hardware Description Language），为了制作数字电路而用来描述 ASIC 和 FPGA 的设计之用。Verilog 的设计者想要以 C 编程语言为基础设计一种语言，可以使工程师比较容易学习。Verilog 是由 Gateway Design Automation 公司于 1984 年开始研发的。Gateway Design Automation 公司于 1990 年被 Cadence Design Systems 公司并购。Cadence 现在对于 Gateway 公司的 Verilog 和 Verilog-XL 模拟器拥有全部的财产权。

2. VHDL：Very-High-Speed Integrated Circuit Hardware Description Language，超高速集成电路硬件描述语言，在基于 CPLD/FPGA 和 ASIC 的数字系统设计中有着广泛的应用。VHDL 语言诞生于 1983 年，1987 年被美国国防部和 IEEE 确定为标准的硬件描述语言。自从 IEEE 发布了 VHDL 的第一个标准版本 IEEE 1076—1987 后，各大 EDA 公司先后推出了自己的支持 VHDL 的 EDA 工具。VHDL 在电子设计行业得到了广泛的认同。此后，IEEE 又先后发布了 IEEE 1076—1993 和 IEEE 1076—2000 版本。

3. Avalon interface：Avalon 总线接口。Nios 系统的全部外设都通过 Avalon 总线与 Nios CPU 相接。Avalon 总线是一种协议较为简单的片内总线，Nios 系统与外界的数据交换通过 Avalon 总线进行。Avalon 总线接口可以分为两类：Master 和 Slave。Master 是一个主控制接口，而 Slave 是一个从控制接口。

## Exercises

1. Keywords

In the article, there are some important words which are the soul of this paper. After reading this paper, we can find some words to stand for this article. Now please find out key words of this paper.

2. Summary

After reading this paper, please write a summarize about PC.

# 科技英语知识 13：动词的翻译

　　同名词一样，英文中动词也存在一词多义的特点，例如，run 的所有词义共计有 200 种以上，每种含义之间既有相近之处，也有一定差别。动词在英文中的使用比较灵活，除了做谓语成分外，还有动名词、现在分词、宾语补足语等用法，很多动词本身还具有名词词性，因此在动词的翻译处理上要从三方面着手：语法成分、语言环境和专业基础，再结合汉语表达习惯和词汇特点，进行透彻理解和灵活表达。

　　例如，work 这个词做动词用时，常常仅被狭义地理解和翻译为"工作"，从而导致译文缺乏文采，生硬难懂。

　　而下列例句中则采用了灵活处理的方式，根据动作的执行主体和客体，对 work 的词义进行引申后，选择汉语中恰当的词来表达。

　　例 1：This method works well.

　　这种方法很有效。

　　句中的 works well 合译为"有效"。

　　例 2：My watch doesn't work.

　　我的表不走了。

　　句中的 works 按汉语习惯译为"走"；也可根据上下文中获得推断信息，把 doesn't work 合译为"坏了"或"停了"。

　　例 3：The machine works smoothly.

　　这台机器运转正常。

　　句中的 works 译为"运转"。

　　work 在其他情况下还可作名词使用，如 idle work power（无功功率）、work frame（框架、机壳）、cable works（电缆厂）、the works of Marx（马克思著作），翻译时要注意区分词性，避免混淆而造成译文的混乱。

# Unit 15　EDA Tools

## Lesson 41　MATLAB

To keep pace with market demand for more performance and functionality in today's mobile phones, digital cameras, computers, automotive systems and other electronics products, manufacturers pack billions of transistors onto a single chip. This massive integration parallels the shift to ever-smaller process geometries, where the chip's transistors and other physical features can be smaller than the wavelength of light used to print them [1]. Designing and manufacturing semiconductor devices with such phenomenal scale, complexity and technological challenges would not be possible without electronic design automation (EDA).

### Introduction to MATLAB

MATLAB is a kind of mathematic software developed by Math Works Company in USA. It is a high level technical language and interactive environment, widely used in algorithm development, data visualization, data analysis, and numerical calculation. It is shown in Figure 41-1 that all the products included in MATLAB, from which you can see the MATLAB consists of two parts: one is MATLAB, and the other is Simulink.

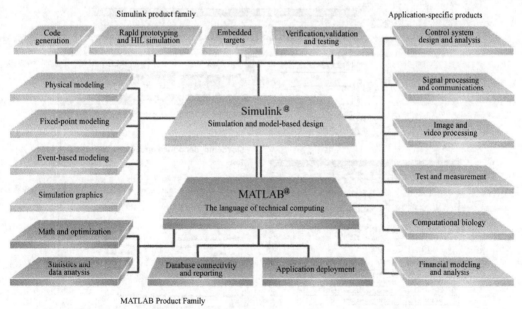

Figure 41-1　MATLAB and Simulink product family

MATLAB is short for Matrix Laboratory. With MATLAB, one can accomplish matrix operations, data and function plotting, the realization of algorithms, creation of user interface, and connection to other programming language procedures. It is mainly used in areas such as engineering calculations, control design, signal processing and communications, image processing, signal detection, financial modeling design and analysis, and so on.

Simulink is an environment for multi-domain simulation and Model-Based Design for dynamic and embedded systems. It provides an interactive graphical environment and a customizable set of block libraries that let you design, simulate, implement, and test a variety of time-varying systems, including communications, controls, signal processing, video processing, and image processing. Add-on products extend Simulink software to multiple modeling domains, as well as provide tools for design, implementation, and verification and validation tasks. Simulink is integrated with MATLAB, providing immediate access to an extensive range of tools that let you develop algorithms, analyze and visualize simulations, create batch processing scripts, customize the modeling environment, and define signal, parameter, and test data. Works in Simulink are shown in Figure 41-2.

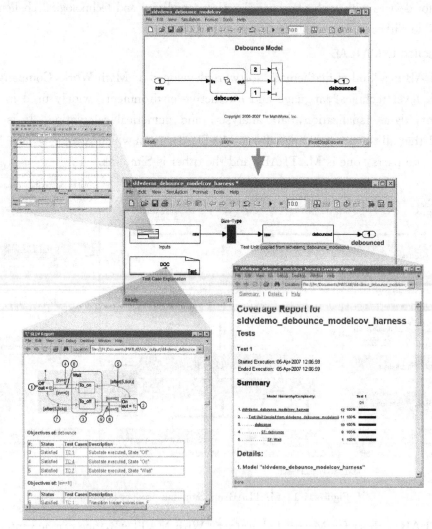

Figure 41-2　Works finished by Simulink

### Getting started with MATLAB

On Windows systems MATLAB is started by double clicking the MATLAB icon on the desktop or by selecting MATLAB from the start menu. The starting procedure takes the user to the Command window where the Command line is indicated with ">>". Used in the calculator mode all MATLAB commands are entered to the command line from the keyboard. MABLAB can be used in a number of different ways or modes: as an advanced calculator in the calculator mode, in a high level programming language mode and as a subroutine called from a C-program.

The MABLAB screen is broken into 3 parts as shown in Figure 41-3. On the right you have the Command Window—this is where you type commands and usually the answers (or error messages) appear here too. On the top left you have the Workspace window—if you define new quantities (called variables) their names should be listed here. On the bottom left you have Command History window—this is where past commands are remembered. If you want to re-run a previous command or to edit it you can drag it from this window to the Command Window to re-run it.

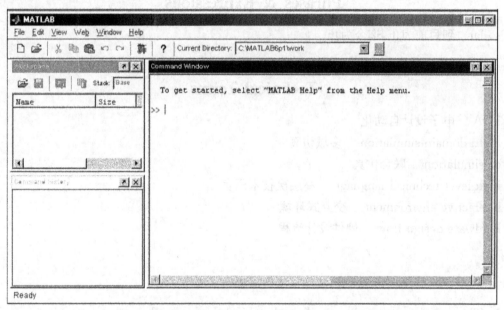

Figure 41-3  The starting interface of MATLAB

### MATLAB and Simulink Link to EDA Simulator

EDA Simulator Link is a co-simulation interface that integrates MATLAB and Simulink into the hardware design flow for the development of application-specific integrated circuits (ASICs) and field programmable gate arrays (FPGAs) [2]. It provides a bidirectional link between MATLAB and Simulink and EDA simulators. EDA Simulator Link software enables you to verify your HDL design from within MATLAB and Simulink. It provides native cosimulation support for both VHDL and Verilog.

So far, the EDA Simulator Link includes the EDA Simulator Link IN, which links

MATLAB to Cadence; the EDA Simulator Link MQ, which connects MATLAB to Mentor Graphics ModelSim; and the EDA Simulator Link DS, which can join the MATLAB and the Synopsys Discovery.

## New Words

algorithm [ˈælgəriðəm] n. 算法
visualization [ˌviʒuəlaiˈzeiʃən] n. 可视化
subroutine [ˌsʌbruːˈtiːn] n. 子程序
bidirectional [ˌbaidiˈrekʃənəl] adj. 双向的
geometry [dʒiˈɔmitri] n. 几何学
modeling [ˈmɔdliŋ] n. [计] 建模，造型
customize [ˈkʌstəmaiz] vt. [计] 定制，用户化

## Phrases & Expressions

so far    到目前为止，迄今为止

## Technical Terms

EDA    电子设计自动化
multi-domain simulation    多域仿真
co-simulation    联合仿真
high level technical language    高层次技术语言
interactive environment    交互式环境
hardware design flow    硬件设计流程

## Notes

1. This massive integration parallels the shift to ever-smaller process geometries, where the chip's transistors and other physical features can be smaller than the wavelength of light used to print them.
   如此庞大的集成度推动了生产工艺向更小的尺寸迈进，芯片上晶体管的尺寸和其他物理特性可以比印制它们的光波长更小。
2. EDA Simulator Link is a co-simulation interface that integrates MATLAB and Simulink into the hardware design flow for the development of application-specific integrated circuits (ASICs) and field programmable gate arrays (FPGAs).
   EDA 仿真器链接是一个联合仿真接口，把 MATLAB 和 Simulink 集成到 ASIC 和 FPGA 的硬件设计流程中。

# Lesson 42　PSpice

SPICE is a powerful general purpose analog circuit simulator that is used to verify circuit designs and to predict the circuit behavior [1]. This is of particular importance for integrated circuits. It was for this reason that SPICE was originally developed at the Electronics Research Laboratory of the University of California, Berkeley (1975), as its name implies: Simulation Program for Integrated Circuit Emphasis.

Many different versions of SPICE are available from many different vendors. SPICE was a text based program. The user was required to describe the circuit using only text, and the simulation results were displayed as text in an output file. Common SPICEs include HSpice, PSpice, and BSpice. SPICE takes a circuit netlist and performs mathematical simulation of the circuit's behavior. A netlist describes the components in the circuit and how they are connected. SPICE can simulate DC operating point, AC response, transient response, and other useful simulations.

PSpice is a PC version of SPICE developed by MicroSim Corp. in 1984 and HSpice is a version that runs on workstations and larger computers. PSpice is available on the PCs in the lab E109.

MicroSim provided a graphical postprocessor, Probe, to plot the results of SPICE simulations. Later, MicroSim, also provided a graphical interface called Schematics that allowed users to describe circuits graphically. The name of the simulation program was changed from PSpice to PSpice A/D when it became possible to simulate circuits that contained both analog and digital devices. MicroSim was acquired by Orcad, which in turn was acquired by Cadence. Orcad improved Schematics and renamed it Capture. "Using PSpice" loosely refers to using Capture, PSpice A/D, and Probe to numerically analyze an electric circuit.

PSpice is a graphical entry version of SPICE, which means that a schematic is drawn on the screen, and the computer derives the NETLIST from the schematic [2]. As a result, PSpice uses the same numerical algorithms that SPICE uses and will still run SPICE files. The computer examines the netlist (or nodes), determines the nodal equations of the circuit and solves them using determinants the same way you solve them in class. The graphical nature of PSpice provides the results on a graph instead of presenting a table of output data as SPICE did.

PSpice can do several types of circuit analyses. Here are some of the most important ones: (1)Non-linear DC analysis: calculates the DC transfer curve; (2)Non-linear transient analysis: calculates the voltage and current as a function of time when a large signal is applied; (3)Linear AC Analysis: calculates the output as a function of frequency. A bode plot is generated; (4)Noise analysis; (5)Sensitivity analysis; (6)Distortion analysis; (7)Fourier analysis; calculates and plots the frequency spectrum; (8)Monte Carlo analysis. In addition, PSpice has analog and digital libraries of standard components (such as NAND, NOR, flip-flops, and other digital gates, op amps, etc). This makes it a useful tool for a wide range of analog and digital applications. All analyses can be done at different temperatures. The default temperature is 300K. The circuit can contain the following components: (1) Independent and dependent voltage and current sources; (2)Resistors; (3)Capacitors; (4) Inductors; (5)Mutual inductors; (6)Transmission lines; (7)Operational amplifiers; (8) Switches; (9)Diodes; (10)Bipolar transistors; (11) MOS transistors; (12)JFET; (13) MESFET; (14)Digital gates (PSpice, version 5.4 and up).

### Getting Started with PSpice Schematics Editor

The first step is to enter your circuit diagram into the computer. You will use Schematics to enter the your circuit.

From the windows desktop or "Start" button, select "Programs", then "PSpice Student " and then "Schematics". You should now be looking at an empty circuit grid similar to the one shown in Figure 42-1. It may not be identical depending on how the previous user left it. You can draw your schematic using the buttons and tools. To complete a schematic, you will be getting the parts of the circuit, placing the parts you selected, connecting the parts to form a circuit, changing the names of the parts, changing the values of the parts, and finally saving the design.

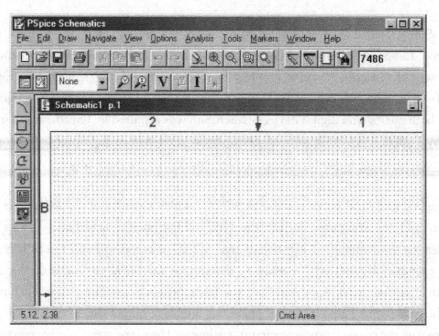

Figure 42-1  PSpice Schamtics window

### Creating an input file for PSpice

When it comes to a relatively simple circuit, you can use the circuit description file only to create a input file for PSpice. Here is an example. A sample circuit is given in Figure 42-2. Draw a schematic of the circuit (on paper), number the nodes and label all elements. Note that the common node (ground) always has number "0". We are interested in all node voltages V1, V2 and all branch currents when the input voltage Vin is equal to 10V.

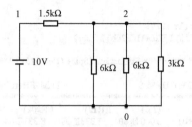

Figure 42-2  A sample circuit

Now you can Create the input file (source file) for PSpice. The source file needs the extension ".cir" in its name. Create the input file with any editor, such as NotePad under windows, etc. Save it on your disk. Save the file as a textfile (.txt when using a word processor such as MS Word). One possible source file may be like this, in Figure 42-3:

```
*EXAMPLE1.CIR
Vs 1 0 10 ;10 V source
R12 1 2 1.5k ;1.5k resistor
R20a 2 0 6k ;6k resistor
R20b 2 0 6k ;6k resistor
R20c 2 0 3k ;3k resistor
.dc Vs list 10 ;DC analysis (Vs=10 V)
.print dc V(1,0) V(2,0) I(R12) I(R20a) I(R20b) I(R20c)
.end
```

Figure 42-3  Source file

### Running the PSpice and displaying the output

From the windows desktop or "Start" button, select "Programs", then "PSpice Student" and then PSpice AD Student. Or you can double click on the filename in your PSpice working directory. Once you are in PSpice, pull down the File menu at the top of the screen and select "Open". The system prompts you for the name of the file. Type in the file name of the circuit you have created before. A window will appear telling you that PSpice program is running, or that the simulation has been completed successfully, or that errors were detected. Click on the "OK" button.

There are two ways to view the output. One method allows you to see the output file. Pull down the File menu, select "Examine Output". The output file contains the input commands, and all the results of the simulations, such as the node voltages, power in the independent voltage sources, current through voltage sources, and the result of the DC sweep as specified in the above source file. In case there was an error, the output also contains a list of errors with a brief explanation. The output file example is shown in Figure 42-4.

```
**********07/01/02 19:32:13 **************Evaluation PSpice (Nov 1999) **************
*EXAMPLE1.CIR
******          CIRCUIT DESCRIPTION
***************************************************************************
Vs 1 0 10 ;10V source
R12 1 2 1.5k ;1.5k resistor
R20a 2 0 6k ;6k resistor
R20b 2 0 6k ;6k resistor
R20c 2 0 3k ;3k resistor
.dc Vs list 10 ;DC analysis (Vs=10V)
.print dc V(1,0) V(2,0) I(R12) I(R20a) I(R20b) I(R20c)
.end
**********07/01/02 19:32:13 **************Evaluation PSpice (Nov 1999) **************
*EXAMPLE1.CIR
******          DC TRANSFER CURVES              TEMPERATURE =   27.000 DEG C
***************************************************************************
 Vs           V(1,0)      V(2,0)       I(R12)      I(R20a)      I(R20b)
 1.000E+01    1.000E+01   5.000E+00    3.333E-03   8.333E-04    8.333E-04
**********07/01/02 19:32:13 **************Evaluation PSpice (Nov 1999) **************
*EXAMPLE1.CIR
******          DC TRANSFER CURVES              TEMPERATURE =   27.000 DEG C
***************************************************************************
 Vs           I(R20c)
 1.000E+01    1.667E-03
            JOB CONCLUDED
            TOTAL JOB TIME                  .08
```

Figure 42-4　Output file

## New Words

verify ['verifai] *v.* 验证
predict [pri'dikt] *v.* 预测
schematic [ski'mætik] *n.* 原理图
workstation ['wəːk steiʃən] *n.* 工作站
transfer [træns'fəː] *v.* 转换

## Phrases & Expressions

in turn　依次
in addition　另外，此外

## Technical Terms

analog circuit　模拟电路
DC　直流
AC　交流
Fourier analysis　傅里叶分析
Monte Carlo analysis　蒙特卡罗分析

**Notes**

1. SPICE is a powerful general purpose analog circuit simulator that is used to verify circuit designs and to predict the circuit behavior.

SPICE 是一个强大的模拟电路仿真器,用于验证电路设计和预测电路的行为。这对于集成电路有着非常重要的意义。

2. PSpice is a graphical entry version of SPICE, which means that a schematic is drawn on the screen, and the computer derives the NETLIST from the schematic.

PSpice 是具有图形输入的 SPICE,电路原理图可以显示在屏幕上,计算机再把原理图的网表提取出来。

# Lesson 43  Cadence

We will precede our lesson about Cadence software with a concise introduction to Cadence Company. Cadence Design Systems is the world's leading EDA Company. Cadence customers use its software, hardware, and services to overcome a range of technical and economic hurdles. Their technologies help customers create mobile devices with longer battery life. Designers of ICs for game consoles and other consumer electronics speed their products to market using its hardware simulators to run software on a "virtual" chip—long before the actual chip exists[1]. The Cadence Company bridges the traditional gap between chip designers and fabrication facilities, so that manufacturing challenges can be addressed early in the design stage. Cadence custom IC design platform enables designers to harmonize the divergent worlds of analog and digital design to create some of the most advanced mixed-signal system on chip (SoC) designs.

Cadence is a large-scale EDA software, with which you can achieve almost all aspects of electronic design, including ASIC design, FPGA design and PCB board design. Cadence has absolute advantage in schematic design, circuit simulation, automatic placement and routing, layout design and verification. Cadence Company has developed her own programming language called Skill[2] and a compiler for Skill language. All of the Cadence tools are written in Skill. The Skill language also provides interfaces with C, so users can develop their own tools based on Cadence.

In this lesson, you will learn about the elements of the Cadence tools.

**1. Full custom IC design tools**

Cadence is known for its full custom IC design ability, which includes tools such as Virtuoso Schematic Composer, Affirma Analog Design Environment, Virtuoso Layout Editor, Affirma Spectra, Virtuoso Layout Synthesizer, Assura Verification Environment,

Dracula and so on. The Cadence software is usually installed in a UNIX or Linux system, which runs on a workstation. Before starting Cadence, we have to log on the Linux system as shown in Figure 43-1.

Figure 43-1  Log on the Linux system

After your login, you can type "icfb &" to start up, as shown in Figure 43-2. The window in Figure 43-2 is called CIW(Command Interpreter Window). CIW is the main user interface of Cadence, which consists of several parts, namely: Title Bar, Menu Banner, Output Area, Input Line, Mouse Bindings Line and Scrolling Bar. Now the Cadence full custom IC design software is initiated, we can complete our work with this marvelous tool.

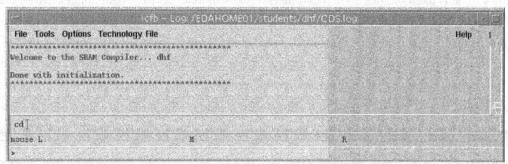

Figure 43-2  Command interpreter window

## 2. Logic design and verification tools

First we come up with an idea, and use the VHDL or verilog HDL to describe the design, creating the HDL codes. Then we do behavioral simulations with the tools such as Verilog-XL, NC-verilog, Leapfrog VHDL, and NC-VHDL etc. to estimate the design, verify the modules function, and debug the project. Next, we employ verisure for Verilog or VHDL Cover for VHDL in the debug and analyze environment to analyze the simulation results. After that we make use of Ambit Build Gates to synthesize, and carry out the gate-level simulations with the SDF files. Finally we execute the fault simulations with verifault [3].

With the simple design flow described above, we go through almost all the tools contained in Cadence LDV. This flow can be fitful for small designs.

### 3. Timing driven DSM design tools

This part is the software for lower level design. The lower level needs iterations. In the design flows before (0.6μm and above), the wire delay is out of consideration, or it is of less influence on designs.

Nowadays, many types of software take the wire delay into account right in the floor plan phase. It is also the demand of the Deep Sub Micron (DSM) design. This is because the wire delay is of the most influence on the whole design, so the influence of this delay in floor plan needs consideration even in the synthesis phase.

In Cadence, there are two main softwares to carry out the timing-driven designs: SE and design planner.

There are many other types of software included in Cadence, such as Board Level Circuit Design Tools and Alta System Level Tools for Wireless Design. With Cadence you can make your dream come true.

## New Words

precede [priː'siːd] *vt.*, *vi.* 开始
harmonize ['hɑːmənaiz] *vt.*, *vi.* 协调
synthesize ['sinθisaiz] *vt.*, *vi.* 综合

## Phrases & Expressions

and so on  等等
come up with  提出，赶上

## Technical Terms

full custom  全定制
behavioral simulation  行为级仿真
mixed signal  混合信号
SoC  片上系统
DSM  深亚微米
wire delay  连线延迟
timing-driven  时序驱动

## Notes

1. Designers of ICs for game consoles and other consumer electronics speed their products to market using its hardware simulators to run software on a "virtual" chip—long before the

actual chip exists.

游戏控制器和其他消费电子的集成电路设计者,在芯片制造之前,应用硬件仿真器进行软件仿真,以加速产品上市的过程。

2. Skill：Cadence 的 Skill 语言是一种基于通用人工智能语言——Lisp 的交互式高级编程语言(Lisp 即 List Processing,表处理)。Skill 语言对使用 Cadence 工具的用户来说,不但可以提高工作效率,而且可以开发用户自己的基于 Cadence 平台的工具。

3. First we come up with an idea, and use the VHDL or verilog HDL to describe the design, creating the HDL codes. ... After that we make use of Ambit Build Gates to synthesize, and carry out the gate-level simulations with the SDF files. Finally we execute the fault simulations with verifault.

首先我们有了一个创意,使用 VHDL 或者 Verilog HDL 来描述设计,创建 HDL 代码；然后应用 Verilog-XL、NC-verilog、Leapfrog VHDL 和 NC-VHDL 工具进行行为仿真、评估设计、验证模块功能、调试工程；接下来使用 verisure 调试 Verilog,或者使用 VHDL Cover 调试 VHDL,分析仿真结果；之后应用 Ambit Build Gates 进行综合,使用 SDF 文件进行门级仿真；最后使用 verifault 进行故障仿真。

# Exercises

### 1. Keywords

In the three lessons, there are some important words which are the soul of the content. After reading the three lessons, we can find some words to stand for each lesson. Now please find out key words of every lesson repectively.

### 2. Summary

After reading this unit, please write a summary of 200~300 words about EDA tools.

# 科技英语知识 14：被动句的翻译

科技文章侧重描述和推理，强调客观准确，所以谓语大量采用被动语态，以避免过多使用第一、二人称而引起主观臆断的印象。英语和汉语都有被动语态，但两种语言在运用和表达上却不尽相同，因此，翻译时必须对被动句做适当的灵活处理。

英语中某些着重被动动作的被动句，为突出其被动意义，可直接译为汉语的被动句，翻译时最常见的方式是在谓语动词前加"被"。但过多地使用同一个词会使译文缺乏文采，因而应根据汉语习惯采用一些其他方式来表达被动语态，如可用"由"、"给"、"受"、"加以"、"把"、"使"等。

例：The machine tools are controlled by PLC.

机床由可编程序控制器控制。

着重描述实物过程、性质和状态的英语被动句，实际上与系表结构很相近，往往可以译成汉语的判断句，即将谓语动词放在"是……的"中间的结构。

例：This kind of device is much needed in the speed-regulating system.

这种装置在调速系统中是很需要的。

英语的被动语态有时可改译成汉语的主动语态。当原句中的主语为无生命的名词，而又无由介词 by 引导的行为主体时，可将原句的主语仍译为主语，按汉语习惯表述成主动句。

汉语中很多情况下，表达被动的含义通常不需要加"被"字，而采用主动语态的形式。如果不顾汉语习惯，强加上"被"字表被动，有时则会使译文看上去不像汉语。

例：The quartz crystal does not vibrate at certain frequency until the voltage is applied.

直到电压加上去以后，石英晶体才会以某一频率振荡。

对于有些被动句，翻译时可将原主语译成宾语，而把原行为主体或相当于行为主体的介词宾语译成主语。

例：Since numerical control was adopted at machine tools, the productivity has been raised greatly.

自从机床采用数控以来，生产率大大提高了。

有些原句中没有行为主体，翻译时可增添适当的主语使译文通顺流畅，如"人们"、"有人"、"大家"、"我们"等。

例：A few years ago it was thought unbelievable that the computer could have so high speed as well so small volume.

几年前，人们还认为计算机能具有如此高的运行速度和如此小的体积是一件难以置信的事。

例：Fuzzy control is found a effective way to control the systems without precise mathematic models.

人们发现，模糊控制是一种控制不具备精确数学模型系统的有效方法。

当不需要或无法讲出动作的发出者时，如表明观点、要求、态度的被动句或描述某地发生、存在、消失了某事物的被动句，可将原句译为汉语的无主句，把原句的主语译作宾语。

例：What kind of device is needed to make the control system simple?

需要什么装置使控制系统简化？

# Unit 16　IC Manuals

## Lesson 44　IC Datasheet

Before IC integrators put chips into use, the IC Datasheets which are usually supplied by the vendors must be carefully read[1]. Every chip has its own datasheet. It is of much benefit for us to know how the IC datasheet is organized. The IC datasheet consists of several parts, such as overview of the chip, the signal description, the functional description, the register description, electrical specifications, mechanical drawings and order information, etc [2].

The chip overview part makes a thorough depiction of the chip, including the chip introduction and the chips of the same series. An example of chip overview is shown in Figure 44-1. It is part of the chip overview of the Yukon 88E8053.

**OVERVIEW**

The single-chip PCI Express based 88E8053 device integrates the Marvell® market-leading Gigabit PHY with the proven Marvell Gigabit MAC and SERDES cores, delivering an ultra-small form factor and high performance. Delivered with the industry's most comprehensive software driver suite, this Yukon device is ideally suited for LAN on motherboard (LOM) and Network Interface Card (NIC) applications. The 88E8053 device is compliant with the PCI Express 1.0a specification. Offered in a 9 x 9 mm, 64-pin QFN package, the 88E8053 reduces board space required for Gigabit LOM implementation significantly.

Figure 44-1　Chip overview of the Yukon 88E8053

Signal description usually includes the input and output signal of the chip. Again, we take the signal description of Yukon 88E8053 for example. Part of the signal description is shown in Figure 44-2.

Functional description contains abundant contents, such as bus type features, memory features, timing, etc. The structure of the functional description is shown in Figure 44-3.

Register description is composed of illuminations of every type of register used in the chip, for example the bus register file, the control register file, etc. The example of Yukon 88E8053 is shown in Figure 44-4.

Electrical specifications cover the electrical information of the IC chip, including the dc characteristics and ac characteristics, etc. An example is shown in Figure 44-5.

Mechanical drawings depict the outline of the IC chip, such as the height, width, etc. A picture of the exterior view of Yukon 88E8053 is shown in Figure 44-6.

Order information part is used for ordering the chip, which often includes the product information and customer information. Part of the order information is shown in Figure 44-7.

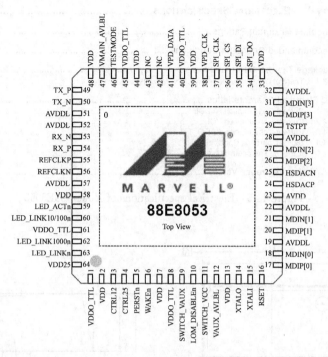

Figure 44-2    Signal description of the Yukon 88E8053

| | | |
|---|---|---|
| Section 2. | Functional Description | 14 |
| 2.1 | Overview | 14 |
| 2.2 | PCI-Express Features | 16 |
| 2.3 | SPI Flash Memory | 18 |
| 2.4 | SPI Flash Memory Loader | 20 |
| 2.5 | TWSI EEPROM | 22 |
| 2.6 | TWSI EEPROM Loader | 23 |
| 2.7 | Plug In Go Unit | 24 |
| 2.8 | Interrupts | 25 |
| | 2.8.1    IRQ Moderation Timer | 27 |
| | 2.8.2    Message Signaled Interrupts (MSI) | 28 |
| 2.9 | Buffer Management Units (BMU) | 28 |
| | 2.9.1    Format of Descriptor and Status List Elements | 28 |

Figure 44-3    Functional description of Yukon 88E8053

| | | |
|---|---|---|
| Section 3. | Register Description | 54 |
| 3.1 | Legend | 54 |
| 3.2 | PCI-Express Configuration Register File | 55 |
| | 3.2.1    Overview and Address Map | 55 |
| | 3.2.2    Registers of PCI Header Region | 58 |
| | 3.2.3    Registers of Header Region | 59 |
| | 3.2.4    Registers of Device Dependent Region | 69 |
| 3.3 | Control Register File | 108 |
| | 3.3.1    Overview and Address Map | 108 |
| | 3.3.2    Registers | 127 |
| 3.4 | GMAC Registers | 213 |
| | 3.4.1    MAC Register Definitions | 213 |

Figure 44-4    Register description of Yukon 88E8053

| Section 4. | Electrical Specifications | 239 |
|---|---|---|
| 4.1 | Absolute Maximum Ratings | 239 |
| 4.2 | Recommended Operating Conditions | 240 |
| 4.3 | Package Thermal Information | 241 |
| | 4.3.1 Thermal Conditions for 64-pin QFN Package | 241 |
| 4.4 | DC Electrical Characteristics | 242 |
| | 4.4.1 Current Consumption AVDDL | 242 |
| | 4.4.2 Current Consumption VDD | 242 |
| | 4.4.3 Current Consumption VDDO_TTL | 243 |
| | 4.4.4 Digital Operating Conditions | 244 |
| | 4.4.5 IEEE DC Transceiver Parameters | 245 |
| 4.5 | AC Timing Reference Values | 246 |

Figure 44-5  Electrical specifications of Yukon 88E8053

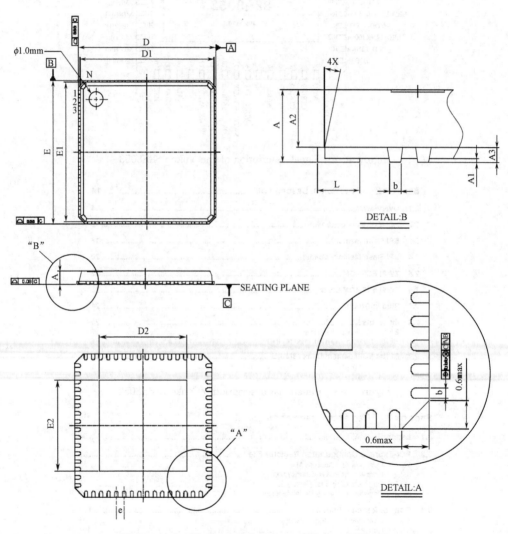

Figure 44-6  Mechanical drawings of Yukon 88E8053

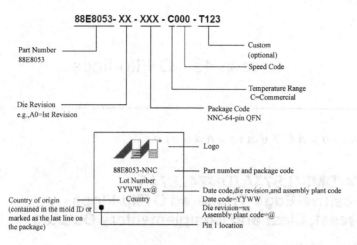

Figure 44-7　Order information of Yukon 88E8053

## New Words

integrator [ˈintigreitə] *n.* 集成者
vendor [ˈvendɔː] *n.* 提供者
depiction [diˈpikʃən] *n.* 描述

## Phrases & Expressions

take … for example　　以……为例

## Technical Terms

IC datasheet　　集成电路数据
signal description　　信号描述
functional description　　功能描述
register description　　寄存器描述

## Notes

1. Before IC integrators put chips into use, the IC Datasheets which are usually supplied by the vendors must be carefully read.
IC 集成者在使用芯片之前，要仔细阅读 IC 提供者提供的数据表。
2. The IC datasheet consists of several parts, such as overview of the chip, the signal description, the functional description, the register description, electrical specifications, mechanical drawings and order information, etc.
IC 数据表包含若干部分，如芯片总揽、信号描述、功能描述、寄存器描述、电气规则、机械制作和订购信息等。

# Lesson 45   D Flip-flops

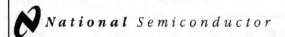

June 1989

## 54LS74/DM54LS74A/DM74LS74A
## Dual Positive-Edge-Triggered D Flip-Flops with Preset, Clear and Complementary Outputs

### General Description

This device contains two independent positive-edge-triggered D flip-flops with complementary outputs. The information on the D input is accepted by the flip-flops on the positive going edge of the clock pulse. The triggering occurs at a voltage level and is not directly related to the transition time of the rising edge of the clock. The data on the D input may be changed while the clock is low or high without affecting the outputs as long as the data setup and hold times are not violated. A low logic level on the preset or clear inputs will set or reset the outputs regardless of the logic levels of the other inputs.

### Features

- Alternate military/aerospace device (54LS74) is available. Contact a National Semiconductor Sales Office/Distributor for specifications.

### Connection Diagram

Dual-In-Line Package

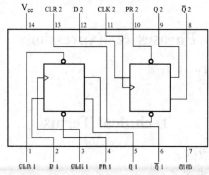

TL/F/6373-1

**Order Number 54LS74DMQB, 54LS74FMQB, 54LS74LMQB, DM54LS74AJ, DM54LS74AW, DM74LS74AM or DM74LS74AN**
See NS Package Number E20A, J14A, M14A, N14A or W14B

### Function Table

| Inputs | | | | Outputs | |
|---|---|---|---|---|---|
| PR | CLR | CLK | D | Q | $\bar{Q}$ |
| L | H | X | X | H | L |
| H | L | X | X | L | H |
| L | L | X | X | H* | H* |
| H | H | ↑ | H | H | L |
| H | H | ↑ | L | L | H |
| H | H | L | X | $Q_0$ | $\bar{Q}_0$ |

H = High Logic Level
X = Either Low or High Logic Level
L = Low Logic Level
↑ = Positive-going Transition
* = This configuration is nonstable; that is, it will not persist when either the preset and/or clear inputs return to their inactive (high) level.
$Q_0$ = The output logic level of Q before the indicated input conditions were established.

## Electrical Characteristics over recommended operating free air temperature range (unless otherwise noted)

| Symbol | Parameter | Conditions | | Min | Typ (Note 1) | Max | Units |
|---|---|---|---|---|---|---|---|
| $V_I$ | Input Clamp Voltage | $V_{CC}$ = Min, $I_I$ = −18 mA | | | | −1.5 | V |
| $V_{OH}$ | High Level Output Voltage | $V_{CC}$ = Min, $I_{OH}$ = Max $V_{IL}$ = Max, $V_{IH}$ = Min | DM54 | 2.5 | 3.4 | | V |
| | | | DM74 | 2.7 | 3.4 | | |
| $V_{OL}$ | Low Level Output Voltage | $V_{CC}$ = Min, $I_{OL}$ = Max $V_{IL}$ = Max, $V_{IH}$ = Min | DM54 | | 0.25 | 0.4 | V |
| | | | DM74 | | 0.35 | 0.5 | |
| | | $I_{OL}$ = 4 mA, $V_{CC}$ = Min | DM74 | | 0.25 | 0.4 | |
| $I_I$ | Input Current @ Max Input Voltage | $V_{CC}$ = Max $V_I$ = 7V | Data | | | 0.1 | mA |
| | | | Clock | | | 0.1 | |
| | | | Preset | | | 0.2 | |
| | | | Clear | | | 0.2 | |
| $I_{IH}$ | High Level Input Current | $V_{CC}$ = Max $V_I$ = 2.7V | Data | | | 20 | μA |
| | | | Clock | | | 20 | |
| | | | Clear | | | 40 | |
| | | | Preset | | | 40 | |
| $I_{IL}$ | Low Level Input Current | $V_{CC}$ = Max $V_I$ = 0.4V | Data | | | −0.4 | mA |
| | | | Clock | | | −0.4 | |
| | | | Preset | | | −0.8 | |
| | | | Clear | | | −0.8 | |
| $I_{OS}$ | Short Circuit Output Current | $V_{CC}$ = Max (Note 2) | DM54 | −20 | | −100 | mA |
| | | | DM74 | −20 | | −100 | |
| $I_{CC}$ | Supply Current | $V_{CC}$ = Max (Note 3) | | | 4 | 8 | mA |

**Note 1:** All typicals are at $V_{CC}$ = 5V, $T_A$ = 25°C.

**Note 2:** Not more than one output should be shorted at a time, and the duration should not exceed one second. For devices with feedback from the outputs, where shorting the outputs to ground may cause the outputs to change logic state an equivalent test may be performed where $V_O$ = 2.25V and 2.125V for DM54 and DM74 series, respectively, with the minimum and maximum limits reduced by one half from their stated values. This is very useful when using automatic test equipment.

**Note 3:** With all outputs open, $I_{CC}$ is measured with CLOCK grounded after setting the Q and $\overline{Q}$ outputs high in turn.

## Switching Characteristics at $V_{CC}$ = 5V and $T_A$ = 25°C (See Section 1 for Test Waveforms and Output Load)

| Symbol | Parameter | From (Input) To (Output) | $R_L$ = 2 kΩ | | | | Units |
|---|---|---|---|---|---|---|---|
| | | | $C_L$ = 15 pF | | $C_L$ = 50 pF | | |
| | | | Min | Max | Min | Max | |
| $f_{MAX}$ | Maximum Clock Frequency | | 25 | | 20 | | MHz |
| $t_{PLH}$ | Propagation Delay Time Low to High Level Output | Clock to Q or $\overline{Q}$ | | 25 | | 35 | ns |
| $t_{PHL}$ | Propagation Delay Time High to Low Level Output | Clock to Q or $\overline{Q}$ | | 30 | | 35 | ns |
| $t_{PLH}$ | Propagation Delay Time Low to High Level Output | Preset to Q | | 25 | | 35 | ns |
| $t_{PHL}$ | Propagation Delay Time High to Low Level Output | Preset to $\overline{Q}$ | | 30 | | 35 | ns |
| $t_{PLH}$ | Propagation Delay Time Low to High Level Output | Clear to $\overline{Q}$ | | 25 | | 35 | ns |
| $t_{PHL}$ | Propagation Delay Time High to Low Level Output | Clear to Q | | 30 | | 35 | ns |

**Absolute Maximum Ratings** (Note)

If Military/Aerospace specified devices are required, please contact the National Semiconductor Sales Office/Distributors for availability and specifications.

| | |
|---|---|
| Supply Voltage | 7V |
| Input Voltage | 7V |
| Operating Free Air Temperature Range | |
|   DM54LS and 54LS | −55°C to +125°C |
|   DM74LS | 0°C to +70°C |
| Storage Temperature Range | −65°C to +150°C |

Note: *The "Absolute Maximum Ratings" are those values beyond which the safety of the device cannot be guaranteed. The device should not be operated at these limits. The parametric values defined in the "Electrical Characteristics" table are not guaranteed at the absolute maximum ratings. The "Recommended Operating Conditions" table will define the conditions for actual device operation.*

## Recommended Operating Conditions

| Symbol | Parameter | | DM54LS74A | | | DM74LS74A | | | Units |
|---|---|---|---|---|---|---|---|---|---|
| | | | Min | Nom | Max | Min | Nom | Max | |
| $V_{CC}$ | Supply Voltage | | 4.5 | 5 | 5.5 | 4.75 | 5 | 5.25 | V |
| $V_{IH}$ | High Level Input Voltage | | 2 | | | 2 | | | V |
| $V_{IL}$ | Low Level Input Voltage | | | | 0.7 | | | 0.8 | V |
| $I_{OH}$ | High Level Output Current | | | | −0.4 | | | −0.4 | mA |
| $I_{OL}$ | Low Level Output Current | | | | 4 | | | 8 | mA |
| $f_{CLK}$ | Clock Frequency (Note 2) | | 0 | | 25 | 0 | | 25 | MHz |
| $f_{CLK}$ | Clock Frequency (Note 3) | | 0 | | 20 | 0 | | 20 | MHz |
| $t_W$ | Pulse Width (Note 2) | Clock High | 18 | | | 18 | | | ns |
| | | Preset Low | 15 | | | 15 | | | |
| | | Clear Low | 15 | | | 15 | | | |
| $t_W$ | Pulse Width (Note 3) | Clock High | 25 | | | 25 | | | ns |
| | | Preset Low | 20 | | | 20 | | | |
| | | Clear Low | 20 | | | 20 | | | |
| $t_{SU}$ | Setup Time (Notes 1 and 2) | | 20 ↑ | | | 20 ↑ | | | ns |
| $t_{SU}$ | Setup Time (Notes 1 and 3) | | 25 ↑ | | | 25 ↑ | | | ns |
| $t_H$ | Hold Time (Note 1 and 4) | | 0 ↑ | | | 0 ↑ | | | ns |
| $T_A$ | Free Air Operating Temperature | | −55 | | 125 | 0 | | 70 | °C |

# Exercises

### Keywords

In this lesson, there are some important words which are the soul of the content. After reading the lesson, we can find some words to stand for the lesson. Now please find out key words of the lesson repectively.

# 科技英语知识 15：说明书常用术语

operational/operating instructions　操作说明书
major components and functions　主要部件及功能
assembles and controls　各部件及其操作机构
group designation　总类名称
operating flow chart　操作流程图
fine adjustment　微调
coarse adjustment　粗调
direction for use　使用方法
wear-life　抗磨损寿命
high voltage cautions　小心高电压
transportation　搬运，运输
instruction for erection　安装规程
power requirements　电源条件
service condition　工作条件
system diagram　系统示意图
wiring/circuit diagram　线路图
operating voltage　工作电压
factory services　工厂检修服务
specific wearability　磨损率
precautions/cautions　注意事项
measuring range　量程
data book　数据表
Don't cast　勿掷
standard accessories　标准附件
accessories supplied　备用附件
safety factor　安全系数
tested error free　经检验无质量问题
ground/GND terminal　接地端子
earth lead　地线
inflammable　易燃物，防火
Keep dry　保持干燥
Keep upright　勿倒置
To be protected from cold/heat　避免遇冷/热
Use roller　（出现在外包装箱上时）移动时使用滚子

warranty 保证书
user's manual 用户手册
features 特点
construction 构造
electric system 电气系统
coolant system 冷却系统
test run 试运转
first commissioning 试车
maintenance 维护，维修
dimensions 尺寸
measurement 尺码
lubrication 润滑
inspection 检验
location 安装位置
fix screw 固定螺钉
specifications 规格
rated load 额定负载
rate capacity 额定容量
nominal speed 额定转速
nominal horsepower 额定马力
Gross/Gr. Wt. 毛重
Net Wt. 净重
fragile 易碎
cutting capacity 加工范围
humidity 湿度
oiling period 加油间隔期
work cycle 工作周期
recyclable 可回收利用的
Handle with care 小心装卸
Heave here 从此提起
haul 起吊，此处起吊
Keep in cool place 置于阴凉处
Not to be tipped 勿倾倒
stuffing 填充料
cleaning 清洗
guarantee 保证书，保修书

# 附　录

## 附录 A　Technical Terms

Science and technology are developing in an amazing high speed. New technical products invade into every corner of people's daily life. Scientific progress greatly promotes the emergence of English technical neologisms (TN). The introduction and development of these neologisms relay on translators' efforts and quality of translation.

1. Technical of terms without hyphen

They are consist of two or more combinations of terms. The front nouns play the role of the adjectives：

vector algebra　矢量代数；

mean square 或 mean-square　均方根；

output error estimation method　输出误差估值法；

organization of data　数据编排；

the collection of information　信息收集。

Combination of terms is also useful component possessive noun：

Newton's Second Law　牛顿第二定律；

Bernoulli's Equation　伯努利方程。

Combination of terms also consist with adjectives and nouns, for example, the following combination is a prefix ＋ adjectives＋nouns：

ultra-high frequency　超高频。

Needs to be noted that the combination of terms which consist with many nouns, Plural nouns also appear in the front or the middle of the situation：

telecommunication transmission engineering　电信传输工程。

Furthermore, some of the singular noun which for "ed" at the end of can also play the role of adjectives：

Wheeled transport　车辆运输；

Increased capacity requirements　增大容量的要求；

Scalar-valued　标量的；

Complex-valued　复值的；

Battery-powered　电池供电的。

2. Combination of terms with hyphen

In many cases, some combination of terms themselves can be used to embellish the other nouns. In order to clear meaning, between the combinations of terms which modification should be add the hyphen.

(1) noun-noun combination

machine code 机器码；machine-code instruction 机器码指令；zero output 零输出；zero-output level 零输出电平。

(2) adjective-noun combination

large screen 大屏幕；large-screen display 大屏幕显示；small quantity 少量；small-quantity production 小批量生产；first generation 第一代；first-generation computer 第一代计算机。

3. Mathematics and mathematical logic in technical English

The pronunciations of the most common mathematical expressions are given in the list below. In general, the shortest versions are preferred (unless greater precision is necessary).

Mathematical symbols:

| | |
|---|---|
| 0 | o (pronunciation: [əu]); naught; zero |
| + | plus (sign); positive sign |
| − | minus (sign); negative sign |
| ± | plus-or-minus sign |
| × | multiplication sign |
| ÷ | division sign |
| = | equality sign |
| ≠ | non-equality sign |
| ≡ | identity sign |
| — | fraction bar |
| / | slant; fraction bar; solidus |
| : | colon; ratio sign, division sign |
| :: | double colon, as (in ratios) (US/Br: a : b :: c : d; Cont. Eu. : a : b = c : d) |
| · | decimal point |
| , | comma |
| ′ | dash, prime (e.g. a′) |
| - | hyphen |
| ∃ | there exists |
| ∀ | for all values of |
| ∼ | tilde |
| √ | root sign; radical sign |
| \| \| | modulus bars |
| ( ) | round brackets; parentheses |
| ( | bracket open |
| ) | bracket close |
| [ ] | (square) brackets |
| { } | braces |

| | |
|---|---|
| # | number; No; item |
| ∞ | infinity sign |
| ≅ | congruence sign |
| ⇒ | implication sign |

Expressing mathematical symbols and terms.

(1) Logic

| | |
|---|---|
| $p \Rightarrow q$ | $p$ implies $q$ / if $p$, then $q$ |
| $p \Leftrightarrow q$ | $p$ if and only if $q$ / $p$ is equivalent to $q$ / $p$ and q are equivalent |

(2) Sets

| | |
|---|---|
| $x \in A$ | $x$ belongs to $A$ / $x$ is an element (or a member) of $A$ |
| $x \notin A$ | $x$ does not belong to $A$ / $x$ is not an element (or a member) of $A$ |
| $A \subset B$ | $A$ is contained in $B$ / $A$ is a subset of $B$ |
| $A \supset B$ | $A$ contains $B$ / $B$ is a subset of $A$ |
| $A \cap B$ | $A$ cap $B$ / $A$ meet $B$ / $A$ intersection $B$ |
| $A \cup B$ | $A$ cup $B$ / $A$ join $B$ / $A$ union $B$ |
| $A \backslash B$ | $A$ minus $B$ / the difference between $A$ and $B$ |
| $A \times B$ | $A$ cross $B$ / the cartesian product of $A$ and $B$ |

(3) Real numbers

| | |
|---|---|
| 1~9 | one to nine; one through nine |
| 1/2 | a (one) half |
| 3/4 | three quarters; three fourths |
| 21/24 | twenty-one twenty-fourths; twenty one over twenty four |
| $3\frac{1}{6}$ | three and one sixth |
| 0.5 | o [əu] point five; zero point five; naught point five |
| .2 | point two |
| 1.0045 | one point zero zero four five |
| $a = b$ | $a$ equals $b$; $a$ is equal to $b$ |
| $a + b$ | $a$ plus $b$ |
| $a - b$ | $a$ minus $b$ |
| $a \pm b$ | $a$ plus or minus $b$ |
| $ab$; $a \times b$ | $a$ times $b$; $a$ multiplied by $b$ |
| $a \div b$ | $a$ divided by $b$ |
| $\frac{a}{b}$ | $a$ over $b$; $a$ divided by $b$ |
| $\frac{ab}{c+d}$ | $a$ times $b$ over $c$ plus $d$ |
| $a/b$ | $a$ slant $b$; $a$ solidus $b$ |
| $a : b$ | $a$ is to $b$ |
| $a : b :: c : d$ | $a$ is to $b$ as $c$ is to $d$ |
| $a \propto b$ | $a$ is proportional to $b$; preferred symbol $a \sim b$ |

| | |
|---|---|
| $a \sim b$ | $a$ is asymptotically equal to $b$; preferred symbol $a \cong b$ |
| $a \approx b$ | $a$ is nearly equal to $b$; $a$ is approximately equal to $b$ |
| $a \equiv b$ | $a$ is identically equal to $b$ |
| $(a-b)(a+b)$ | $a$ minus $b$, $a$ plus $b$ |
| $a \neq 5$ | $a$ (is) not equal to 5 |
| $\therefore a=b$ | therefore $a$ equals $b$ |
| $\because a=b$ | because $a$ equals $b$ |
| $a \to b$ | $a$ tends to $b$; $a$ approaches $b$ |
| $a!$ | factorial $a$; $a$ factorial |
| $a < b$ | $a$ is less than $b$; $a$ is smaller than $b$ |
| $a > b$ | $a$ is greater than $b$; $a$ is larger than $b$ |
| $a \ll b$ | $a$ is much less than $b$; $a$ is much smaller than $b$ |
| $a \gg b$ | $a$ is much greater than $b$, $a$ is much larger than $b$ |
| $a \geq b$ | $a$ is larger than or equal to $b$ |
| $0 < a < 1$ | zero is less than $a$ is less than 1 |
| $0 \leq a \leq 1$ | zero is less than or equal to $a$ is less than or equal to 1 |
| $a_b$ | $a$ subscript $b$; $a$ sub $b$ |
| $a'$ | $a$ dash, $a$ prime |
| $a''$ | $a$ double dash; $a$ double prime; $a$ second prime |
| $a^2$ | $a$ squared |
| $a^3$ | $a$ cubed; $a$ to the third power |
| $a^4$ | $a$ to the fourth; $a$ to the power four |
| $a^n$ | $a$ raised to the power $n$ |
| $a^{-n}$ | $a$ to the minus $n$th power |
| $a^{bx}$ | $a$ to the power $bx$ |
| $a^{\frac{1}{4}}$ | $a$ to the power one-fourth |
| $(a+b)^2$ | $a$ plus $b$ all squared |
| $\left(\dfrac{a}{b}\right)^2$ | $a$ over $b$ all squared |
| $\sqrt{a}$ | square root of $a$ |
| $\sqrt[3]{a}$ | cube root of $a$ |
| $\sqrt[n]{a}$ | $n$th root of $a$ |
| $\sqrt{a^2+b^2}$ | square root out of $a$ squared plus $b$ squared |
| $|a|$ | modulus of $a$; absolute value of $a$; magnitude of $a$ |
| $\begin{vmatrix} a_{11} & a_{12} \\ a_{21} & a_{22} \end{vmatrix}$ | determinant (matrix) |

  first row: a sub one one, a sub one two,
  second row: a sub two one, a sub two two

| | |
|---|---|
| $\binom{n}{k}$ | binominal $n$ over $k$ |
| $\sum_{a=1}^{n}$ | the sum from $a$ equals one to $n$ |
| $\prod_{a=1}^{n}$ | the product from $a$ equals one to $n$ |
| $\sin \alpha$ | sine $\alpha$; sine of $\alpha$ |
| $\cos \alpha$ | cosine $\alpha$; cosine of $\alpha$ |
| $\tan \alpha$ | tangent $\alpha$; tangent of $\alpha$ |
| $\cot \alpha$ | cotangent of $\alpha$ |
| $\sec \alpha$ | secant $\alpha$; secant of $\alpha$ |
| $\sinh \alpha$ | shine cos $\alpha$; hyperbolic sine of $\alpha$ |
| $\cosh \alpha$ | cosh cos $\alpha$; hyperbolic cosine of $\alpha$ |
| $\operatorname{sech} \alpha$ | sech $\alpha$; hyperbolic secant of $\alpha$ |
| $\arccos x$ | inverse cosine $x$; cos minus one $x$ |
| $\lg x$ | lg of $x$ |
| $\log_a x$ | logarithm to the base $a$ of $x$ |
| $d$ | differentiation sign |
| $dx/dy$ | $dx$ by $dy$ (derivative of $x$ with respect to $y$) |
| $\bar{x}$ | first derivative of $x$ with respect to time |
| $\bar{x}'$ | second derivative of $x$ with respect to time |
| $\delta x$ | variation of $x$; delta $x$ |
| $\delta x/\delta a$ | partial derivative of $x$ with respect to $a$ |
| $\lim f(x)$ | limit of the function of $x$ |
| $\int_a^b f(x)dx$ | definite integral of $f(x)$ from $x = a$ to $x = b$ |
| $\iint$ | double integral |
| $\iiint$ | triple integral |
| $\oint$ | circuital integral; integral round a closed circuit |
| $\int f(x)dx$ | indefinite integral of $f(x)dx$ |

(4) Linear algebra

| | |
|---|---|
| $\|x\|$ | the norm (or modulus) of $x$ |
| $\overrightarrow{OA}$ | vector $\mathbf{OA}$ |
| $\overline{OA}$ | the length of the segment $OA$ |
| $\mathbf{A}^T$ | transpose matrix of $\mathbf{A}$ |
| $\mathbf{A}^{-1}$ | inverse of the square matrix $\mathbf{A}$ |

# 附录 B  Scientific English Writing

## Introduction to Scientific English

Scientific English falls into two parts: the scientific papers and books and non-scientific papers and books. The former part takes the uniform style in the structures and expressions, while the latter part is of diversity in structures, even in the vocabularies and sentences.

The scientific papers include a few components, namely: title, author's name, abstract, introduction, body, acknowledgement, appendix, references and resume. The structure of scientific books is almost the same as papers. Although scientific books have different lengths, their contents are organized clearly with different levels. The contents are arranged by **Volume**, **Part**, **Chapter and Section**. Other scientific writings cover product specifications, technical materials, patents, technical contracts, experiment instructions, and so on. Experiment instructions comprise the equipments introductions, the experimenting steps and pre-readings as well as the instructions. Patent literatures have a uniform format in writing, in which some of the words are barely used in the scientific papers and books, such as **hereby**, **thereby**, **whereby**, etc. Besides, patents in different countries differ from each others. Technical contracts have some characters same as that of patent literature.

## Organizing the Scientific Writing

In general, scientific writings consist of several main parts, namely: the beginning, the text, the ending, the index, and the reference. In this session, you will learn something about organization of the scientific writing.

In the beginning of the scientific books and periodicals or articles, the purpose of the writing must be explained to readers. You can propose some problems right at the beginning. For example, you may find these expressions useful: *In this chapter*, *we begin with a review of* ..., *we discuss some measures associated with* ..., *we develop some methods for handling* ..., which begins with someone; and *This paper*, *presents an analysis of* ..., *describes various designs for* ..., *derives some important algorithms that can be used in* ..., which are leaded by the paper or chapter; and *An algorithm for* ... *is proposed*, *A new recursive algorithm for* ... *is introduced*, *based on* ..., *A new recursive algorithm for* ... *is developed*, whose subject are the content of the writing, and *This paper is concerned with* ..., *is devoted to analyze* ..., *is intended to describe*..., whose subject are the article or chapters.

In some articles, especially the books, the language is usually very concise to express the characters. You may use expression to supply the readers with the principles or practical knowledge, for example:

*The intent is to provide the reader with a sound understanding of the fundamentals involved in analyzing, planning, designing, and evaluating... It also provides a solid base for further study and research.*

You may also find these examples to be useful:

The goal of this book is to bridge the gap existing between classical and existing control theory.

There are some dictions to express the content of the articles or books:

*The material covered in this book is broad enough to satisfy a variety of backgrounds and interests, thereby allowing considerable flexibility in making up the course material.*

In the end of every chapter of the articles and books there are summary or conclusions to draw a whole picture of the content for readers. The style of writing summary or conclusions is the same with the beginning.

## Expressions about mathematical content

When you encounter something about maths, you have to refer to the vocabulary of mathematics. Some of these words will be used under fixed conditions. For example, **Solution** will be used instead of *solve* when you are solving, and **Proof** will be used instead of *prove* when you are proving.

Mathematic nomenclature named after someone takes four kind of forms, for instance: **Laplace transform**, **Fourier transform**, **Schwarz inequality**, which are names of someone used as attributes; **Euler' theorem**, **Jordan's lemma**, which are names of someone followed by a "'"; **Markovian process**, **Guassian distribution**, which are names of someone transformed to adjectives; **Runge-Kutta method**, **Guass-Jocobi climination method**, which are names of two or more people connected with a "-".

Expressions for operations are usually constructed by simple sentences or phrases. Common sentence structures include participle phrases and gerunds. Participle phrases are composed of verbs dealing with the operations. The main clause of this kind often takes "*we have (obtain, get, find)* " as predicate, whose object is the result of the operation. The subject in the gerund style is the gerund transformed from the verb related to the operation, and the predicate is always the verbs such as *give(s)*, *yield(s)*, *get(s)*, *become(s)*, and so on.

## On Data Format and Unit in Scientific English

It is known to all that in scientific English there are lots of figures, tables, formulas, symbols, units, numbers, abbreviations, and punctuations, which can called the data and format. In this section some necessary information will be provided to help the ones who may have some problems about data and format in scientific English.

There are many styles of line, such as solid line, dash line, dotted line, heavy line, light line, dot-dash line, shaded line, piecewise linear line and so on. You can choose any one of them to form a curve in a picture. Words like graph, plot, drawing, diagram, view, chart, sketch, map, etc. can be used to different kinds of pictures. In one paper there can possibly be only one picture, so you have to number them in order, for example: Figure 1, or Figure 1-1, etc. The number of figure is always followed by the name, which gives a concise explanation of the content. If the pictures are not created by the author, it is important to give clear indication of the source.

Another format of data is table. There are usually several rows and columns in a table, where you can fill your data according to the name of rows and columns. Tables are numbered in order just as same as figures, for example: Table 1, or Table 1-1, etc. The difference is that the titles of figure lie on bottom of figures, while that of table always lie on top.

There are many forms of numerals in English, such as common numbers, exact numbers, approximate numbers, etc. Common numbers in scientific English can be straightly written in Arabic numerals, for example: 1, 100, 46 958, etc. You can use words like total, most, least, etc. to express the exact number of something, and words like about, around, some, etc. to express the approximate number of something. It is necessary to point out that a sentence can't be started with a number, for example:

"**80 students are in the classroom.**" (**Wrong**)

You can use the following sentence instead:

"There are 80 students in the classroom." (Right)

In the scientific writing for international communications, the SI is often adopted. The SI is divided into two series: one is primary unit system, and the other is derived unit system. Some of the units are named after somebody, and others are composed of a single symbol or combination of symbols.

## Scientific English Expressions Usage for Writing

There are many differences between Chinese and English expressions, especially in scientific field. In this section you will learn something about scientific English expressions usage for writing a paper.

When writing in scientific English, it always happens to you that you have to define something or explain them, especially for some nouns or terminology. You can use "define" or "definition" directly to express the exact meaning of something. For example:

**Definition: The difference between the largest and smallest class boundaries is called the range of the table.**

You can also use passive voice of "define". For instance: "M is defined as N." You may also use "refer to as" or "call" or "know as" to make a definition. For example:

"M is called/ known as/ referred to as N."

In scientific English, there are lots of expressions about time from ancient times to modern times. To express a date, American English will put the month before the day, with the year at the end. For example: "March 12th, 2009." British English will be a little different, in which the day is put in front of the month, with the year at the end. For example: "12th March, 2009." Besides, you have to pay attention to the use of the prepositions such as in, on, at, etc.

When things in your paper must be classified, you may refer to the words type, class, sort, variety, category, etc. All the stuffs to be sorted can be written in one paragraph, even in one sentence. For example:

**There are two kinds of electric charge in the world. One is positive charge, and the other is negative charge.**

It is often the case that in your scientific researches you have to compare your result with the others'. In scientific English, there are many ways to express things of various extents. You can use the positive degree, comparative degree, superlative degree of the adjectives to reflect them. Or you can also use different words to depict them. For example, you may refer to something as "useful", "more useful", and "most useful"; or you can use "very useful" and "extremely useful" instead.

To express cause and effect there are two types of sentences: one takes the cause as the theme; the other the effect. You can use "because" or a clause before or after the main clause to express the cause of something, while you have to put the words to express the effect, for example "therefore", right in the main clause. In English, words to express the cause and effect never meet in one sentence.

## Introduction to Scientific Paper Contribution

Besides contributing to the domestic conferences or journals, we inevitably submit our scientific researches to international conferences or periodicals. This is one of the most important academic intercourses among the world. The papers required strictly by international conferences or journals bring too much difficult for the beginners. There are fixed steps from contributing to publishing. In this section you will learn something about contribution to international conferences and periodicals.

Generally, the international conferences will be hold by international scientific organization every year at different places all over the world. Before the conference there will be a call for papers, which gives clear indication of submission deadline, format of the papers, etc. An example of this is shown in Figure 1. We should read it carefully before submission. If we want our paper to be accepted, we have to comply with the requirement of the conference, such as submitting before deadline, novel content, etc. There are two forms of communication: one is oral presentation, the other is poster.

Figure 1   Example of call for papers

There is something different between periodical contribution and conference contribution. The basic difference lies in that papers for periodicals are the whole article, not summary. The periodicals held by IEEE are most influential. Just like the call for papers of conference, there is "information for authors", which usually includes the content requirement, format requirement, etc., in every kind of periodical. An example of this is shown in Figure 2.

Apart from the papers submitted, you have to write a letter to the reviewers and a resume about yourself.

Figure 2　Information for authors

# 参 考 文 献

[1] Tarmo Anttalainen. Introduction to Telecommunications Network Engineering[M]. Second Edition. British Library Cataloguing in Publication Data, 2003.

[2] M Rafiquzzaman. Fundamentals of Digital Logic and Microcomputer Design[M]. Fifth Edition. John Wiley & Sons, Inc. , 2005.

[3] Alberto Leon-Garcia, Indra Widjaja. Communication Networks Fundamental Concepts and Key Architectures[M]. McGraw-Hill, 2001.

[4] Electronic Instrument Handbook[M]. McGraw-Hill, 2004.

[5] Jon Orwant, Jarkko Hietaniemi, John Macdonald. MasteringAlgorithms with Perl[M]. O'Reilly & Associates, Inc. , 1999.

[6] K R Rao, Zoran S Bojkovic, Dragorad A Milovanovic. Introduction to Multimedia Communications Applications, Middleware, Networking[M]. Johe Wiley & Sons, Inc. , 2006.

[7] Luís Correia. Mobile Broadband Multimedia Networks Techniques, Models and Tools for 4G[M]. Jordan Hill, 2006.

[8] Mooichoo Chuah, Qinqing Zhang. Design and Performance of 3G Wireless Networks and Wireless Lans[M]. Springer, 2006.

[9] Yamaguchi Noriyasu. Advanced Automated Fingerprint Identification System[J]. IEEE, 1998.

[10] Darren Ashby, Bonnie Baker. Circuit Design, Elsevier,2008.

[11] John Bird. Electrical Circuit Theory and Technology[M]. Second Edition. Elsevier Science, 2003.

[12] Umeshk Mishra, Jasprit Singh. semiconductor device physics and design[M]. Springer.

[13] Sheng S Li. Semiconductor physics electronics[M]. Second Edition. Springer.

[14] John Bird. Electrical Circuit Theory and Technology[M]. Second Edition. Elsevier Science, 2003.

[15] Pong P Chu. RTL Hardware Design Using VHDL Coding for Efficiency, Portability, and Scalability[M]. John Wiley & Sons, Inc. , 2006.

[16] John B Casey, Ken Aupperle. Digital Television and the PC[M]. Hauppauge Computer Works, Inc. , 1998.

[17] Horowitz Paul, Winfield Hill. The Art of Electronics[M]. Second Ed. Cambridge University Press, 1989.

[18] Sears Francis W, Mark W Zemansky. Hugh D Young. University Physics[M]. Sixth Ed. Addison-Wesley Publishing Co. , 1982.

[19] Donald A Neamen. 半导体器件导论[M]. 北京:清华大学出版社,2006.
[20] Akira Sekiguchi, Rei Kinjo, Masaaki Ichinohe. A Stable Compound Containing a Silicon-Silicon Triple Bond[J]. Science,2004,Vol. 305, No. 5691:1755~1757.
[21] O'Mara, William C. Handbook of semiconductor Silicon Technology. William Andrew Inc., 1990:349~352.
[22] Paul Horowitz, Winfield Hill. The Art of Electronics[M]. Cambridge University Press, 1989.